Gynecologic and Obstetric Investigation

Founded 1895 as 'Monatsschrift für Geburtshilfe und Gynäkologie',
continued 1946–1969 as 'Gynaecologia' and 1970–1977 as 'Gynecologic Investigation'

Founders: A. Martin and M. Sänger
Former Editors: E. Anderes (1939–1952), Th. Koller (1939–1969),
O. Käser (1954–1969), R. Wenner (1959–1967), P. Bloch (1965–1969),
W.L. Hermann (1970–1975), P.J. Keller (1976–1984), G. Zador (1985–2001)

KARGER

Printed in Switzerland
on acid-free paper by
Reinhardt Druck, Basel

Appears 6-weekly:
2 volumes per year
(8 issues)

Guidelines for Authors

Gynecologic and
Obstetric Investigation

Submission

Only original papers, carefully selected case reports, and invited reviews written in English are considered and should be sent to:

Editorial Office 'Gynecologic and Obstetric Investigation'
S. Karger AG
P.O. Box
CH–4009 Basel (Switzerland)
Tel. +41 61 306 1344
Fax +41 61 306 1434
E-Mail s.aeschbach@karger.ch

Mailing address for courier deliveries only:
Allschwilerstr. 10
CH–4055 Basel (Switzerland)

Manuscripts should be submitted in quadruplicate (with four sets of illustrations of which one is an original), typewritten double-spaced on one side of the paper, with a wide margin and on a 3.5-inch disk. The preferred word processing package is Word for Windows®. Presentation of manuscripts should conform with the Uniform Requirements for Manuscripts Submitted to Biomedical Journals (see N Engl J Med 1997;336:309–315).

Conditions

All manuscripts are subject to editorial review. *Original Papers* should not exceed 4 printed pages (12 manuscript pages), including tables, illustrations and references. *Articles on current investigation* are limited to 2 printed pages (6 manuscript pages). *Case Reports* are carefully selected and limited to 2 printed pages (6 manuscript pages): this includes statements of the problem, patient history, diagnosis, treatment, results and any other information pertinent to the case and maximum 5 references. *Reviews* are usually commissioned; however, the Editor-in-Chief is happy to discuss potential articles with authors who would like to contribute. Submission of an article for publication implies the transfer of the copyright from the author to the publisher upon acceptance. Accepted papers become the permanent property of 'Gynecologic and Obstetric Investigation' and may not be reproduced by any means, in whole or in part, without the written consent of the publisher. It is the author's responsibility to obtain permission to reproduce illustrations, tables, etc. from other publications.

All manuscripts must be accompanied by a covering letter signed by all authors. Assurance should be given that the manuscript is not under simultaneous consideration by any other publication.

All manuscripts originating from non-English-speaking countries must be revised by a professional linguistic reviewer and it must be evident from the covering letter that this has been done.

Good clinical practice: It has become mandatory that every trial in humans must first obtain approval from an independent Ethics Committee and formal, informed consent from the patients before they participate in a clinical study or experiment. In order to avoid unnecessary delay with the review of manuscripts, authors are kindly asked to state, preferably in the Materials and Methods section, that approval and informed consent have been obtained. These two statements must also appear on the covering letter which accompanies every manuscript and is signed by each author.

Arrangement

All pages should be consecutively numbered beginning with the title page, then the text, acknowledgements, references and legends to figures. The text in original papers should be divided under the headings: Abstract, Introduction, Material(s) and Method(s), Results, and Discussion. The text in case reports should be divided under the headings: Introduction, Case Report(s) and Conclusion(s).

Title page: The first page of each paper should indicate the title, the authors' names, the institute where the work was conducted, and a short title for use as running head.

Full address: The exact postal address of the corresponding author complete with postal code must be given at the bottom of the title page. Please also supply phone and fax numbers, as well as e-mail address.

Key words: For indexing purposes, a list of 3–10 key words in English is essential.

Abstract: The first page of the text should include an abstract of up to 10 lines.

Footnotes: Avoid footnotes. When essential, they are numbered consecutively and typed at the foot of the appropriate page.

Tables and illustrations: Tables and illustrations (both numbered in Arabic numerals) should be prepared on separate sheets. Tables require a heading and figures a legend, also prepared on a separate sheet. For the reproduction of illustrations, only good drawings and original photographs can be accepted; negatives or photocopies cannot be used. Due to technical reasons, figures with a screen background should not be submitted. When possible, group several illustrations on one block for reproduction (max. size 180 × 223 mm) or provide crop marks. On the back of each illustration, indicate its number, the author's name, and 'top' with a soft pencil. Electronically submitted b/w halftone and color illustrations must have a final resolution of 300 dpi after scaling, line drawings one of 800–1200 dpi.

Letters to the Editor: This section is set aside for critical comments directed to a specific article that has been published in the journal. Letters should be brief (not exceeding 500 words), double spaced and limited to a maximum of 5 citations. The letters and replies should be prepared according to journal format. Illustrative material is only permitted with permission of the Editor-in-Chief. With your correspondence, please include your complete mailing address, telephone and fax numbers, and email addresses, if available. The Editor-in-Chief reserves the right to refuse letters, shorten letters, delete objectional comments, and make other changes to to comply with the style of the journal. Send all Letters to the Editor to the above address.

Color illustrations: Up to 6 color illustrations per page can be integrated within the text at the special price of CHF 660.–/USD 545.00 per page. Color illustrations are reproduced at the author's expense.

References: In the text identify references by Arabic numerals [in square brackets]. Material submitted for publication but not yet accepted should be noted as 'unpublished data' and not be included in the reference list. The list of references should include only those publications which are cited in the text. Do not alphabetize; number references in the order in which they are first mentioned in the text. The surnames of the authors followed by initials should be given. There should be no punctuation other than a comma to separate the authors. Preferably, please cite all authors. Abbreviate journal names according to the Index Medicus system. (Also see International Committee of Medical Journal Editors: Uniform requirements for manuscripts submitted to biomedical journals. N Engl J Med 1997;336:309–315.)

Digital Object Identifier (DOI)
S. Karger Publishers supports DOIs as unique identifiers for articles. A DOI number will be printed on the title page of each article. DOIs can be useful in the future for identifying and citing articles published online without volume or issue information. More information can be found at www.doi.org.

Examples
(a) Papers published in periodicals: Sun J, Koto H, Chung KF: Interaction of ozone and allergen challenges on bronchial responsiveness and inflammation in sensitised guinea pigs. Int Arch Allergy Immunol 1997;112:191–195.
(b) Papers published only with DOI numbers: Theoharides TC, Boucher W, Spear K: Serum interleukin-6 reflects disease severity and osteoporosis in mastocytosis patients. Int Arch Allergy Immunol DOI: 10.1159/000063858.
(c) Monographs: Matthews DE, Farewell VT: Using and Understanding Medical Statistics, ed 3, revised. Basel, Karger, 1996.
(d) Edited books: Parren PWHI, Burton DR: Antibodies against HIV-1 from phage display libraries: Mapping of an immune response and progress towards antiviral immunotherapy; in Capra JD (ed): Antibody Engineering. Chem Immunol. Basel, Karger, 1997, vol 65, pp 18–56.

Page charges

There are no page charges for papers of 4 or fewer printed pages (including tables, illustrations and references). Each additional complete or partial page is charged to the author at CHF 275.–/USD 196.00. The allotted size of a paper is equal to approx. 12 manuscript pages (including tables, illustrations and references).

Galley proofs

Unless indicated otherwise, galley proofs are sent to the first-named author and should be returned with the least possible delay. Alterations made in galley proofs, other than the correction of printer's errors, are charged to the author. No page proofs are supplied.

Reprints

Order forms and a price list are sent with the galley proofs. Orders submitted after the issue is printed are subject to considerably higher prices.

KARGER

Fax +41 61 306 12 34
E-Mail karger@karger.ch
www.karger.com

The Guidelines for Authors are available at:
www.karger.com/goi_Guidelines

Gynecologic and
Obstetric Investigation

General Information

Publication data: 'Gynecologic and Obstetric Investigation' is published 8 times a year. Volumes 57–58, each with 4 issues, appear in 2004.

Subscription rates: Subscriptions run for a full calendar year. Prices are given per year.
Personal subscription:

Print or Online	Print/Online combined
CHF 815.–	CHF 897.–
EUR 604.–	EUR 664.–
USD 652.00	USD 718.00
postage and handling	(added to print and print/online)
CHF 42.40 Europe,	CHF 62.80 Overseas
EUR 30.40	
USD 42.40	

Institutional subscription:

Print or Online	Print/Online combined
CHF 1630.–	CHF 1760.–
EUR 1208.–	EUR 1304.–
USD 1304.00	USD 1408.00
postage and handling	(added to print and print/online)
CHF 52.80 Europe,	CHF 78.40 Overseas
EUR 38.–	
USD 52.80	

Airmail surcharge: CHF 52.80 / USD 44.00
Discount subscription prices:
Society for Gynecologic Investigation

Microforms: Please direct all enquiries and orders to:
University Microfilms Inc.
300 North Zeeb Road, P.O. Box 1346
Ann Arbor, MI 48106-1346 (USA)

Subscription orders: Orders can be placed at agencies, bookstores, directly with the Publisher
S. Karger AG, P.O. Box
CH–4009 Basel (Switzerland)
or with any of the following distributors:
Deutschland: S. Karger GmbH, Postfach
D–79095 Freiburg
France: Librairie Luginbühl
36, bd de Latour-Maubourg
F–75007 Paris
Great Britain, Ireland: S. Karger AG
4 Rickett Street
London SW6 1RU, UK
USA: S. Karger Publishers, Inc.
26 West Avon Road, P.O. Box 529
Farmington, CT 06085, USA
India, Bangladesh, Sri Lanka:
Panther Publishers Private Ltd.
33, First Main
Koramangala First Block
Bangalore 560 034, India
Japan: Karger Japan, Inc.
Yushima S Bld. 3F
4-2-3, Yushima, Bunkyo-ku
Tokyo 113-0034, Japan

Change of address: Both old and new address should be sent to the subscription source.

Single issues and back volumes: Information can be obtained through the Publisher.

Advertising: Correspondence and rate requests should be addressed to the Publisher.

Bibliographic indices: This journal is regularly listed in bibliographic services, including *Current Contents®* and Index Medicus/MEDLINE.

Book Reviews: Books are accepted for review by special agreement.

Copying: This journal has been registered with the Copyright Clearance Center (CCC), as indicated by the code appearing on the first page of each article. For readers in the US, this code signals consent for copying of articles for personal or internal use, or for the personal or internal use of specific clients, provided that the stated fee is paid per copy directly to
Copyright Clearance Center Inc.
222 Rosewood Drive
Danvers, MA 01923 (USA)
A copy of the first page of the article must accompany payment. Consent does not extend to copying for general distribution, for promotion, for creating new works, or for resale. In these cases, specific written permission must be obtained from the copyright owner,
S. Karger AG,
P.O. Box, CH–4009 Basel (Switzerland).

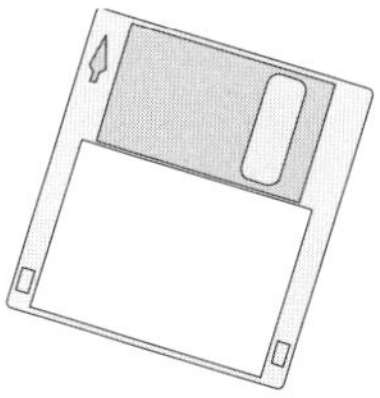

Disk Submission

A **comprehensive version** of these guidelines for the **submission of manuscripts and illustrations on disk** can be accessed online under **www.karger.com/disksubmission** or requested as a leaflet from the editorial department.

Please follow the general rules on style and arrangement in the **'Guidelines for Authors'** of the journal. Submit a **double-spaced printout** of your manuscript matching the file on disk exactly.

Technical Information
Store only data pertinent to this manuscript. Include your fonts (screen and printer fonts for Macintosh). For a **text only** manuscript, use a 3.5-inch disk. For **text and illustrations,** use 3.5 or 5.25 Magneto Optical Disc; CD-ROM (DOS or hybrid); iomega Zip/DOS (100 MB), or 44 MB, 88 MB or 200 MB SyQuest. Compress extensive data using ZIP format.

Data Description
Label your disk with the **title of the journal or book** for which it is submitted; a **short version** of the paper's title; your **name,** postal and e-mail address, fax and phone number; **hard- and software** used, including the version; and the **complete list of the file names.** File names consist of eight characters before the dot identifying the file and three characters after the dot identifying the application.

Text and Format
The **preferred word-processing package is MS-Word.** Other commonly used PC text programs and ASCII formats are accepted.
Update your data when you revise subsequent to the reviewer's or editor's decision and mark revisions clearly. Manuscripts accepted for publication are subject to **copy editing:** changes and/or questions will be marked in the manuscript, which will be returned to you with your disk and the galley proofs for checking.
Enter your text continuously flush left. Do not use any layout functions. Do not split words at the end of a line. **Do not use indentations.** Use a blank line before and a hard return at the end of paragraphs.
Arrange headings flush left. Insert only **one space after punctuation marks.** Insert Greek letters, mathematical symbols etc. with your word-processing program. Use **boldface and italics** as well as sub- and superscript where appropriate.
Tables are part of the text and should be placed at the end of the text file. Separate the individual parts of a table by lines. Use the tabulator instead of spaces. Format operation data on one line in the same column.

Legends to illustrations are part of the text and should also be listed at the end of the text file.

Illustrations
Store illustrations as separate files: do not integrate them in your text. Export **line drawings/vector graphics** as TIF, EPS or WMF format. Use Photoshop for processing and retouching **scanned half-tone images.** Save the original scan and the processed version. Export **black and white or color images** as TIF or EPS format in their anticipated size in print. **For detailed information** on how to submit illustration data on disk please consult the leaflet or **www.karger.com/disksubmission.**

KI 02 DISK

KARGER

Fax +41 61 306 12 34
E-Mail karger@karger.ch
www.karger.com

The Journal Home Page is available at:
www.karger.com/goi

The Fetus as a Patient 2004

Please visit our website at http://www.macc.jp/fetus2004

Fukuoka, Nippon
April 23 (Friday) - April 26 (Monday), 2004
Invited by Hitoo Nakano

Congress Venue: Sea Hawk Hotel & Resort
2-2-3 Jigyohama, Chuo-ku, Fukuoka, 810-8650, JAPAN

Registration Fee	Registration before December 25, 2003	Registration on or after December 25 and On-Site
Active Participant	JPY 40,000	JPY 50,000
Midwife & Co-Medical	JPY 25,000	JPY 35,000
Accompanying person	JPY 20,000	JPY 25,000
Banquet (April 24, 2004)	JPY 10,000	JPY 10,000

Abstract Submission: January 30, 2004

Tentative Scientific Program

April 23, 2004

Ian Donald School 1
Opening Welcome

April 24, 2004

Plenary Session 1-3
Symposium 1 Diagnosis of structural defects 1
Symposium 2 Diagnosis of structural defects 2
Symposium 3 Screening in early pregnancy 1
Symposium 4 Screening in early pregnancy 2
Symposium 5 Fetal / neonatal cardiology and circulation
Poster Presentation 1
Ian Donald School 2
Hakata Night

April 25, 2004

Plenary Session 4 -7
Symposium 6 Fetal cell, gene and ES cell transplantation
Symposium 7 Fetal growth and hypoxia 1
Symposium 8 Fetal growth and hypoxia 2
Symposium 9 Brain damage (fetus / neonate)
Symposium 10 New technology 1
Symposium 11 New technology 2
Poster Presentation 2
Ian Donald School 3
Congress Dinner

April 26, 2004

Plenary Session 8-9
Symposium 12 Fetal / neonatal neurology
Symposium 13 Fetal Therapy: surgery and medication
Symposium 14 Ethics and social problems in fetal medicine
Symposium 15 Maternal complications
Symposium 16 Preterm labor
Closing

F 04220

Congress Secretariat:
c/o MA Convention Consulting Inc.
Dai 2 Izumi-shoji Bldg., 4-2-6 Kojimachi, Chiyoda-ku,
Tokyo 102-0083, JAPAN
Phone:+81-3-5275-1191 Fax:+81-5275-1192
E-mail: info@macc.jp

Gynecologic and
Obstetric Investigation

Gynecol Obstet Invest 2004;57:1–60
DOI: 10.1159/000075234

Published online: December 29, 2003

Short Papers

The Baboon as a Nonhuman Primate Model for the Study of Human Reproduction

Editors
D'Hooghe, T.M., Leuven
Mwenda, J.M., Nairobi
Hill, J.A., Boston, Mass.

KARGER

Basel · Freiburg · Paris · London · New York ·
Bangalore · Bangkok · Singapore · Tokyo · Sydney

The publication of these short papers was supported by:

ESHRE (European Society for Human Reproduction and Embryology)

WHO (World Health Organization)

ICRO-UNESCO (International Cell Research Organization)

SERONO International

National Center for Research in Reproduction, Nairobi, Kenya

Institute of Primate Research, Nairobi, Kenya

Leuven University Fertility Center, Dept. Obstetrics and Gynecology, University Hospital Gasthuisberg, Leuven, Belgium

S. Karger
Medical and Scientific Publishers
Basel · Freiburg · Paris · London
New York · Bangalore · Bangkok
Singapore · Tokyo · Sydney

Drug Dosage
The authors and the publisher have exerted every effort to ensure that drug selection and dosage set forth in this text are in accord with current recommendations and practice at the time of publication. However, in view of ongoing research, changes in government regulations, and the constant flow of information relating to drug therapy and drug reactions, the reader is urged to check the package insert for each drug for any change in indications and dosage and for added warnings and precautions. This is particularly important when the recommended agent is a new and/or infrequently employed drug.

P.O. Box, CH–4009 Basel (Switzerland)
Printed in Switzerland on acid-free paper by
Reinhardt Druck, Basel
ISBN 3–8055–7588–2

KARGER

Fax + 41 61 306 12 34
E-Mail karger@karger.ch
www.karger.com

Vol. 57, No. 1, 2004

Gynecologic and
Obstetric Investigation

Contents

KARGER

Fax + 41 61 306 12 34
E-Mail karger@karger.ch
www.karger.com

Access to full text and tables of contents, including tentative ones for forthcoming issues: www.karger.com/goi_issues

Endometriosis: Baboon Model and in vitro Models

Infertility Issues in the Developing World

International Strategy for Funding Reproductive Biomedical Research in Primates

1

A Critical Review of the Use and Application of the Baboon as a Model for Research in Women's Reproductive Health

T.M. D'Hooghe[a, b], *J.A. Hill*[c], *J.M. Mwenda*[a]

[a]Institute of Primate Research, Karen, Nairobi, Kenya; [b]Leuven University Fertility Center, Department of Obstetrics and Gynecology, University Hospital Gasthuisberg, Leuven, Belgium; [c]The Fertility Center of New England, Reading, MA, USA

Women's Reproductive Health

Over the last years there has been increasing medical, political and social awareness regarding various aspects of Women' s Reproductive Health including prevention of unwanted pregnancy, prevention of obstetric and neonatal morbidity and mortality, prevention and treatment of infertility, sexual problems, specific aspects of pregnancy-induced hypertension (PIH), atherosclerosis, cancer, toxicology, drug abuse. This increased awareness can partly be explained by the progress in medical knowledge in the area of reproductive health and sexuality, but also by the knowledge that women respond differently from men in many aspects of pharmacokinetics, brain function, cardiovascular diseases, auto-immune diseases, cancer. At least as important is the social and political awareness that women are different but equal to men, and have a unique role in raising children, keeping families together and promoting social and economic development in many parts of the world. As a result, there is increasing interest in gender studies, sexuality issues, development projects coordinated by women in the South, and research institutes dealing with Women's Health, like the Department of Women's Health at the National Institutes of Health (NIH, USA).

Contribution of Primate Research to Reproductive Health in Women

A detailed understanding of human reproduction and reproductive disorders is not easy due to ethical restrictions. All biomedical issues related to conception and early human life are bound to be scrutinized and analyzed critically, not only by physicians or scientists, but also by ethicists, sociologists, politicians and the media. In a pluralistic world, it is hard to find broad consensus in matters such as abortion, embryo manipulation, genetic therapy, therapeutic cloning, non-genetic parenthood, etc. In contrast, it is generally accepted in biomedicine and by the public opinion that new diagnostic or therapeutic methods in reproductive medicine should be safe and efficient before application in humans. This assessment of safety and efficiency is difficult since human reproduction is a unique biological process, fundamentally different from reproduction in rats, mice, rabbits or even pigs, goats or cattle. Only nonhuman primates like the great apes (chimpanzees, gorilla, urang-utans), baboons and rhesus monkeys, are in most aspects similar to humans in terms of reproductive anatomy and physiology. Therefore, it is not surprising that some reproductive diseases, like endometriosis, only occur in nonhuman primates and not in other animals. Clearly, nonhuman primates may be the only animal models that are really relevant for research in human reproduction. The great apes are endangered species and can only be used for research if there are extremely important medical reasons, i.e. development of vaccines against HIV to prevent AIDS. Although baboons are not endangered, they in fact represent a threat to the agriculture of many Sub-Saharan African countries, and are killed in large numbers by African farmers. Therefore, it is not unlogical to consider the baboon as a research model for human reproduction, since this is a medically and ethically very important process that can not be studied in a clinical meaningful way in other animal models.

The Need for Basic and Preclinical Research in Nonhuman Primates before Application of New Assisted Reproductive Technology in Humans

In the area of assisted reproductive technology (ART), the need for primate models is increasing due to ethical constraints of reproductive research in humans and to increased public awareness about the safety of new developments in reproductive technology. Two world experts in ART, Schatten [1] and Winston and Hardy [2] have recently highlighted serious concerns about safety issues in reproductive medicine. According to Schatten, innovations in clinical ART are introduced swiftly, typically without animal studies, and the accuracy of extrapolations from research models remain unclear [1]. Winston and Hardy [2] also stress that 'Some therapies, which on assessment are highly empirical or unproven, are being used in human subjects before they have been validated by proper cell culture experiments or detailed animal research' [2]. There is concern about the efficiency and the safety for nearly all new technologies that are being developed or already applied in humans, especially regarding genomic imprinting and premature exposure of gametes and embryos to potentially damaging growth factors from in vitro culture media. The health of children born after these techniques and their impact on future generations is of course very important. These concerns are related to intracytoplasmatic injection of immature or abnormal sperm, spermatogonial stem cell transplantation, in vitro oogenesis and maturation, embryogenesis after cryopreservation and thawing of immature and mature oocytes, development of new culture media for long term culture and their effect on embryo quality, clinical application of embryonic stem cells, therapeutic cloning. The practice of cytoplasmic transfer for the treatment of 'infertility related to cytoplasmatic dysfunction in older women' has been introduced in clinical practice without significant tests regarding safety and efficiency. Women with recurrent IVF failure received small amount of cytoplasm containing mitochondria from young and healthy donor oocytes. As a result, both donor and receptor mtDNA could be passed on to the next generations and these children could be exposed to mitochondrial diseases. An unusually high rate of Turner Syndrome has been found in children born after cytoplasmic transfer [3]. These risks have prompted the FDA to seriously limit this practice to well designed clinical trials and to promote preclinical research in relevant animal models like nonhuman primates. For all these issues, a preclinical model is necessary after research in vitro and/or on small animal models and before clinical application in humans.

Why Are Primate Models Not Fully Utilized for Biomedical Research in Reproductive Health?

Many researchers often claim that primate research is inaccessible and unavailable for research due to the lack of significant primate colonies in their countries, to the high cost related to primate research or to ethical aspects related to primate research. However, it is important to realize that there are several primate centers in the source countries such as Kenya, where these primates can be housed in conditions that mimick their natural habitat and where research costs are lower. Of course this cannot be an excuse for unethical research. Primate research can only be justified as preclinical research for serious human medical conditions or for biomedical interventions without proven efficiency or safety in humans that may deeply affect the health of future generations, when research on other animal models or in vitro is either irrelevant or not sufficiently relevant to make important scientific progress. In the area of assisted reproductive technology, this would imply mainly research on primate gametes and primate embryos and assess the effect of various interventions on fetal and neonatal development, and health of primate offspring over several generations. In this context, it would rather be unethical to develop new techniques in mice and apply them in humans without efficiency and safety testing in nonhuman primates.

Furthermore, scientific collaboration with primate centers in the South could represent a new dimension of North-South collaboration and could promote the development of Centers of Excellence in the South, which could maintain a win-win relationship with research institutions from the North.

The Institute of Primate Research, Nairobi, Kenya and the Baboon as a Well-Characterized Model for Research in Human Reproduction

The Institute of Primate Research (IPR) in Nairobi, Kenya, is a biomedical research institute in sub-Saharan Africa recognized as 'WHO Collaborating Center', and uses internationally accepted guidelines for research in nonhuman primates. One of the main interests of IPR is the area of reproduction, besides infectious diseases, ecology and conservation. Contraception has been a major area of interest for many years. The baboon model for endometriosis has been developed at IPR since the last 12 years and it is clear that baboons offer many advantages for reproductive research when compared to other nonhuman primates [4]. More recently, research projects have been initiated at IPR regarding assisted reproduction and embryo implantation. In this issue of Gynecologic and Obstetric Investigation, the reader can find a selection of invited short papers prepared for the Serono Symposia International Conference on 'The baboon as a model for research in human reproduction', scheduled to take place in Nairobi, Kenya, 11–14th January 2004. The aim of this meeting was to review how baboons have been and can be used as research models for basic and clinical aspects of human reproduction and how they can be considered as a preclinical model for new diagnostic and therapeutic approaches in reproductive medicine. Unfortunately, this conference had to be cancelled due to international terrorist threats in Nairobi. Obviously, we are very disappointed about this situation, but still plan to hold the meeting in another format at another place in the future. However, we decided to proceed with the publication of this special issue of Gynecologic and Obstetric Investigation. We hope that its contents may contribute to increase the awareness in the scientific community and in the society that preclinical primate research in the area of reproduction can not only be justified but is essential to demonstrate the efficacy and safety of new treatments, on the condition that the indications for research are well selected and the welfare of the study subjects has been assured.

References

1 Schatten GP: Safeguarding ART. Nature Cell Biol 2002;4(suppl):19–22.
2 Winston RML, Hardy K: Are we ignoring potential dangers of IVF and related treatments? Nature Cell Biol 2002;4(suppl):14–18.
3 Cohen J, Scott R, Schimmel T, Levron J, Willadsen S: Birth of infant after transfer of anucleate donor oocyte cytoplasm into recipient eggs. Lancet 1997;350:186–187
4 D'Hooghe TM, Cuneo S, Debrock S, Chai C, Mwenda JM: Baboon model for fundamental and preclinical research in endometriosis. Gynecol Obstet Invest 2004;57:43–47.

2

Need for Primate Models in Biomedical Research

J.L. VandeBerg

Southwest National Primate Research Center and Department of Genetics, Southwest Foundation for Biomedical Research, San Antonio, TX, USA

Introduction

What organism is most appropriate for biomedical research aimed at understanding the biology of humans in healthy and diseased states? Although the answer depends in part on the nature of the question being asked, it often is pluralistic. Advances in understanding human biology are dependent on results derived from a collage of animal species, ranging from insects and fishes all the way up to humans themselves.

Obviously, the best species to use from the standpoint of relevance to the human condition is humans themselves; and indeed, a major portion of biomedical research is conducted with human subjects. However, many experimental manipulations cannot be conducted ethically with human subjects, others would be outlandishly expensive to conduct with human subjects, and still others would require an unacceptably long duration of time to conduct with humans.

For these reasons, it is necessary to use animal models. Selection of an animal model involves compromises between conflicting objectives. On one hand, it is desirable for animal models to share many genetic biochemical, physiological, and anatomical characteristics with humans so that experimental results are highly relevant to the human condition. Species with close phylogenetic relationships to humans not only share many biological characteristics, but also have many of the same genes influencing relevant phenotypes functioning on a similar genetic background. In addition, it is desirable to be able to impose controlled environmental conditions (e.g., agent, dose, duration, etc.) that are as similar as possible to those experienced by humans. The mechanisms of action of environmental agents should resemble those known or suspected in humans, because a frequent objective in animal model research is to characterize in detail the biological mechanisms influencing the trait of interest. These considerations indicate that nonhuman primates are optimal models for research on complex physiological processes and diseases that have significant genetic and environmental components.

On the other hand, constraints of time, money, physical facilities, and availability of animal species are frequently in conflict with the requirements for an 'ideal' animal model. Other considerations being equal, investigators would prefer to complete experiments and obtain answers in weeks or months rather than years. It would be desirable to use small animals in order to minimize caging, feeding, and other care costs.

When the human system under investigation is so basic that it has major similarities with homologous systems in all vertebrates or all mammals, then duration of time and expense required to answer the question can be minimized by the use of invertebrates or cold-blooded vertebrates. Thus, the vinegar fly, *Drosophila melanogaster,* has been a favored model for research on fundamental principles of animal genetics for nearly a century, and the zebrafish has become a favored model for research on basic vertebrate developmental processes.

However, invertebrates and cold-blooded vertebrates are not sufficiently similar to humans for research on many aspects of anatomy, physiology, behavior, and genetics. For research on systems that have characteristics shared only by mammals, rodents are the species of choice in regard to money, space, and time required to conduct the research. Thus, the laboratory mouse is the most widely used model for research on mammalian genetics, and the laboratory rat is the most widely used model for research on mammalian physiology.

However, their short life spans and major differences between rodents and humans in genetic and physiological characteristics make rodents unacceptable for some purposes. Furthermore, many anatomical, physiological, behavioral, and genetic characteristics of humans have features that are shared only with other primates. These distinct features interact with each other and with environmental factors in complex ways that are unique to primate species. For research on these complex features and their interactions, only nonhuman primates can serve as informative models.

Nonhuman Primates as Models for Common Multifactorial Diseases

Common multifactorial diseases, such as cardiovascular disease, hypertension, diabetes, Alzheimer's, and Parkinson's, are the predominant cause of morbidity and mortality in industrialized countries. How will develop one or another of the common diseases is determined by a large number of genetic factors and a large number of environmental factors, all of which interact with one another. Some of the genetic factors and some of the environmental factors are unique to primates.

Although non-primate models have made and will continue to make important contributions to our understanding of multifactorial diseases, they do so by modeling one or a few of the multitude of the factors that contribute to these diseases. They cannot model the entirety of the genetic, physiological, and environmental universe of factors that interact in determining human susceptibility to the development and progression of these diseases.

For example, consider atherosclerosis as one of the most common multifactorial diseases. The mouse is a useful model for investigating some aspects of this disease, but it is dissimilar from humans in many factors known to be critical to this disease. As examples, the mouse is principally herbivorous whereas humans share with many other primates a diet that has substantial amounts of animal fat and cholesterol; the mouse does not share with humans a social structure that can be well modeled in regard to human-like stress, which is a risk factor of atherosclerosis; and the mouse genome is lacking some of the genes that are critical to human cholesterol metabolism, and which have genetic variants known to be risk factors of atherosclerosis.

The advent of transgenic technologies has enabled the transfer of some of these human genes into mice where they can be studied one at a time, or in combinations of two or three, on the genetic background of the mouse. But mice cannot replace primate models for research that requires a close correspondence to human subjects in the complex interactions of dozens of genes, on a physiological background that is determined by thousands of genes. In addition, because of the phylogenetic and physiological disparity between rodents and humans, some genes that are important determinants of atherosclerosis could not possibly be discovered from research with mice, e.g., *LPA* and *CETP*, although rodent models with these human transgenes have been developed. For these reasons, nonhuman primates must be included in the collage of animals used to model atherosclerosis and other complex human diseases if we are going to develop an understanding of their complex etiologies.

We have established the baboon, *Papio hamadryas*, as a model for research on the complex genetic and dietary factors that affect risk of atherosclerosis in humans, and on the interactions among them. Baboons resemble humans in lipemic response to an atherogenic diet, relatively low cholesterol absorption and turnover rates, contribution of bile acid and neutral sterol to total fecal excretion of cholesterol, and changes in excretion of cholesterol when fed an atherogenic diet [1–3]. In addition, they have naturally occurring arterial lesions that resemble those of humans [4], and the two species have similar gene maps that can be closely aligned [5]. It is constellations of complex characteristics such as these that make nonhuman primates essential species among the collage of animal models for multifactorial human diseases.

Some recent advances from research on risk factors of atherosclerosis in the baboon model include the localization to specific chromosomal regions of gene loci that significantly affect sodium-lithium countertransport which is strongly associated with blood pressure [6], fat-free mass as a measure of adiposity [7], plasma LDL cholesterol which is associated with risk of atherosclerosis [8], plasma HDL cholesterol which is inversely related to risk of atherosclerosis [9], as well as other physiological risk factors of multifactorial diseases. In addition, we have used the baboon model to develop a novel positional cloning strategy for identifying the responsible genes much more rapidly than is possible with conventional approaches [9]. Although some of the responsible genes may be genes that already had been identified from research with human subjects, we anticipate that others will be novel genes that are currently unknown. These discoveries will create new opportunities for understanding genetic effects on the etiology of atherosclerosis and for developing new strategies for the prevention and treatment of this common disease.

Our group is conducting a parallel study with human subjects [10], but the baboon model provides several critical advantages that enhance the power of genetic analysis. Baboons can be maintained long term on identical diets and subjected to carefully controlled dietary challenges, eliminating diet as confounding variable when searching for genes that affect physiological traits. Other environmental factors that contribute to variability in human populations also can be carefully controlled in studies with baboons. In addition, baboons can be bred to produce large pedigrees of extended families that are ideal for genetic analysis; for example, we have produced sire families containing more than 100 half and full sibs. Finally, baboons can be subjected to experimental procedures that are not ethical with human subjects, for example serial liver biopsies while being fed a series of diets to investigate changes in gene expression in response to specific dietary components [9].

Nonhuman Primates as Models for Infectious Diseases

The need for nonhuman primate models for research on infectious diseases is also obvious. Ethical and legal considerations mandate that we develop and test drugs and vaccines for treating infectious diseases using animal models rather than human subjects. As is the case for multifactorial diseases, rodents and other non-primate species serve as important models for research on infectious diseases. However, susceptibility to infection by particular pathogens, characteristics of the immune response, and consequent characteristics of disease progression vary greatly among the mammalian orders.

Host specificity of human hepatitis viruses has required the use of chimpanzees in research on the development of drugs and vaccines for hepatitis B and hepatitis C. Only the apes share with humans susceptibility to infection with these viruses. It is known that some humans who become infected with hepatitis C virus (HCV) spontaneously clear the virus, whereas others become chronic carriers. In an exciting experiment to determine what host genes affect an individual's ability to clear HCV, Bigger et al. [11] took serial liver biopsies from an experimentally infected chimpanzee and applied microarray technology to investigate changes in gene expression. They observed a major change in interferon response gene expression, correlated in time with viral clearance. This example illustrates the value of the chimpanzee model for research that could not be conducted with humans or with other animals.

AIDS is another infectious disease for which nonhuman primates are indispensable models. Chimpanzees, which unlike monkeys can be infected with HIV, have been used to screen large numbers of candidate vaccines so that the few that appeared to have efficacy could be subjected efficiently to human trials [12]. The development of the rhesus macaque model of AIDS, which closely resembles human AIDS in its pathophysiology, has enabled rapid progress in developing vaccine strategies for preventing disease progression in individuals who already are infected.

And Chagas' disease, caused by the parasite *Trypanosoma cruzi* with which an estimated 16–18 million people are infected [13], has a very different course of pathophysiological development in mice than in primates. Although mice have been used for many years as models to investigate the effects of infection with *T. cruzi*, primate models that mimic the human manifestations of this disease are now being developed [14]. There exist no prophylactics or vaccines for Chagas' disease and no effective drug therapies for chronic carriers of the parasite; and as many as half of infected persons die of cardiac disease, or megaesophagus or megacolon. Therefore, the addition of nonhuman primates to the collage of animal models used in research on Chagas' disease has great potential for developing preventions and treatments.

Conclusion

Nonhuman primates also are critical models for humans in many other areas of biomedical research, such as reproduction, development, and behavior and behavioral disorders. The rationale for using nonhuman primates is much the same regardless of the system being modeled: human subjects are not practical for some of the experimental approaches involved, and non-primates are too dissimilar from humans in some of the characteristics under investigation. Because humans share with other primate species so many anatomical, physiological, behavioral, and genetic characteristics, nonhuman primates will continue to serve a critical role in many biomedical research arenas in the years ahead.

Supported by NIH grants P01 HL28972, P51 RR13986, and R01 RR16347.

References

1 Dell RB, Mott GE, Jackson EM, Ramakrishnan R, Carey KD, McGill HC Jr, Goodman DS: Whole body and tissue cholesterol turnover in the baboon. J Lipid Res 1985;26:327–337.
2 Eggen DA: Cholesterol metabolism in rhesus monkey, squirrel monkey, and baboon. J Lipid Res 1974;15:139–145.
3 Mott GE, Jackson EM, Morris MD: Cholesterol absorption in baboons. J Lipid Res 1980;21:635–641.
4 McGill HC Jr, Strong JP, Holman RL, Werthessen NT: Arterial lesions in the Kenya baboon. Circ Res 1960;8:670–679.
5 Rogers J, Mahaney MC, Witte SM, Nair S, Newman D, Wedel S, Rodriguez LA, Rice KS, Slifer SH, Perelygin A, Slifer M, Palladino-Negro P, Newman T, Chambers K, Joslyn G, Parry P, Morin PA: A genetic linkage map of the baboon *(Papio hamadryas)* genome based on human microsatellite polymorphisms. Genomics 2000;67:237–247.
6 Kammerer CM, Cox LA, Mahaney MC, Rogers J, Shade RE: Sodium-lithium countertransport activity is linked to chromosome 5 in baboons. Hypertension 2001;37:398–402.
7 Comuzzie AG, Martin LJ, Cole SA, Rogers J, Mahaney MC, Blangero J, VandeBerg JL: A quantitative trait locus for fat free mass in baboons localizes to a region homologous to human chromosome 6. Obes Res 2001;9 (suppl 3):S71.
8 Kammerer CM, Rainwater DL, Cox LA, Schneider JL, Mahaney MC, Rogers J, VandeBerg JL: Locus controlling LDL cholesterol response to dietary cholesterol is on baboon homologue of human chromosome 6. Arterioscler Thromb Vasc Biol 2002;22:1720–1725.
9 Cox LA, Birnbaum S, VandeBerg JL: Identification of candidate genes regulating HDL cholesterol using a chromosomal region expression array. Genome Res 2002;12:1693–1702.
10 MacCluer JW, Stern MP, Almasy L, Atwood LA, Blangero J, Comuzzie AG, Dyke B, Haffner SM, Henkel RD, Hixson JE, Kammerer CM, Mahaney MC, Mitchell BD, Rainwater DL, Samollow PB, Sharp RM, VandeBerg JL, Williams J: Genetics of atherosclerosis risk factors in Mexican Americans. Nutr Rev 1999;57:S59–65.
11 Bigger CB, Brasky KM, Lanford RE: DNA microarray analysis of chimpanzee liver during acute resolving hepatitis C virus. J Virol 2001;75: 7059–7066.
12 Murthy KK, Cobb EK, Rouse SR, Lunceford SM, Johnson DE, Galvan AR: Correlates of protective immunity against HIV-1 infection in immunized chimpanzees. Immunol Lett 1996;51:121–124.
13 WHO: Chagas Disease: Tropical Diseases, Tropical Disease research Progress, 1991–1992. Eleventh Programme Report of the UNDP/World Bank/ WHO Special Programme for Research and Training in Tropical Diseases. Geneva, World Health Organization, 1993, pp 67–75.
14 Zabalgoitia M, Ventura J, Anderson L, Carey KD, Williams JT, VandeBerg JL: Morphologic and functional characterization of Chagasic heart disease in non-human primates. Am J Trop Med Hyg 2003;68:248–252.

3
Science, Ethics and Regulation in Primate Research

J.P. Hearn

Research School of Biological Sciences, Australian National University, Canberra, Australia

Introduction

The past 50 years has seen an enormous advance in our understanding of primate biology. The study of non-human primate systems at all levels from molecular, systemic, social and environmental is often the closest that we can come to investigate the fundamental factors that influence human biology. During the same period, the ability to use non-invasive technologies has accelerated our capacity for sequential studies at ever more precise levels of investigation.

It is not over dramatic to say that most of us would be dead or debilitated if it were not for the advances made through primate research into normal and disease processes. While the flow of benefit appears at first sight to favour the human primate, it is also true to say that many advances in human medicine may flow back to influence the improved management and conservation of captive and wild primates.

In this presentation, I submit that there is a virtuous circle in the balance of research opportunities, ethical constraints and regulatory instruments that can provide advances in knowledge and benefits for all primates.

Research Opportunities

Increasingly, advances in basic research can translate rapidly into advances in medicine and in other strategic fields with major benefits for human and animal wellbeing. As indicated above, there is an increasing trend towards studies being interdisciplinary, non-invasive and requiring fewer numbers of animals to deliver significant results because of the precision of the experimental tools. In many instances, however, one cannot escape from the need for invasive procedures when the benefits justify them and for the breeding, captive management and euthanasia of animals when this is essential.

Examples of such imperatives include the study of AIDS and the underlying questions in immunology that effect this and other infectious diseases that still scourge billions of humans. Fundamental questions in genetics, developmental biology, stem cell biology, teratology and toxicology, cardiology, surgery, neurobiology and ageing are a few examples where studies of varying degrees of invasion may be necessary to obtain results. The contemporary fields of genomics, proteomics and phenomics provide some of the most exciting opportunities in bio-medical history. The monitoring of emerging diseases and their interactions between animals including primates and humans holds special interest.

A feature of many programs is the close integration of molecular, cellular and physiological study with the behavioural, ecological and conservation dimensions that can provide a more holistic approach. It is in some of these areas that research carried out in primate source countries such as Kenya can have special impact. Almost all biomedical research is carried out on less than ten primate species, almost all of which are not endangered or threatened. Indeed several are pests in their home countries and are widely eradicated when they come in conflict with human populations. In such cases, the captive breeding of stocks derived from such species can provide a major research result that not only delivers new knowledge and therapies but can also facilitate and enhance the participation of scientists from less developed countries in the global research effort to address some of the diseases referred to above – which often are major inhibitors for the development of those same countries. It would be unethical not to use such assets in these circumstances.

Ethical Aspects

Non-human primates are by definition our closest genetic relatives. In studying the similarities and differences, a stringent ethical approach is essential. The related debate on animal research and 'rights' is not the focus here but it is essential that the debate is facilitated, the procedures and ethical dimensions are transparent and that there is careful consideration and consensus in the resulting regulations. Independent ethical viewpoints are essential and need to be represented in the debate.

At the heart of such debate and diologue is the respect for all life and our appropriate valuation of life at all levels in permitting a sustained balance for survival of the biota and therefore of the planet. Whenever the issues are over-simplified the debate can become non-productive, while the dilemmas attached to advances in research resulting in non-sustainable increases in human population challenge our ability to deliver a balance that benefits both humans and animals.

A further feature in the ethical debate is the differences between cultures, religions and regional views towards the use of animals in research and in conservation, both in captivity and in the wild. Education and training programs are essential to expose varied opinions and the reasons for them. In some instances, principles that derive from widely different premises cannot be reconciled. In such cases, it is important to review the related ethical debates around the world and to strive for best practice and for respect.

There are several well-regarded publications that present ethical guidelines for the care and use of primates in research; as well as similar guidelines for some field studies. These guidelines, such as those of the International Primatological Society; the Institute for Laboratory Animal Resources of the US National Academy of Sciences; and those of the Universities Federation of Animal Welfare (UK) provide well thought out and deeply considered approaches that attempt to resolve the complex dilemmas referred to above. In some cases however, these guidelines require an additional consideration when research is being carried out in primate source countries. The climatic, nutritional, veterinary, housing, management, breeding and experimental conditions and facilities may require quite different approaches to those in more developed countries that are not the source of primate populations and therefore need to create wholly independent environments and conditions.

In achieving best practice, it is essential that the available knowledge and resources be used to best advantage. Primates are special species and are usually available only in scarce numbers. The research is usually long-term and costly in every respect. Wherever possible new communications technologies must be used to share samples, analysis, facilities and other major resources. There are several important international assets that can help such coordination, teamwork and conservation. Among these are the Primate Infonet and the World Directory of Primatologists, both managed from the Wisconsin National Primate Research Centre in Madison. These, easily accessed databases provide immediate information on the research disciplines of most of the world's primatologists, research laboratories and field sites. They were established to enhance communication and to encourage collaboration and conservation of primate resources in captivity and in the wild. For reference, please go to http://www.primate.wisc.edu/

Regulation

Many countries insist on stringent regulation of all aspects of primate management, conservation and research. Indeed, the array of legislative and regulatory guidelines, instructions or orders that have multiplied in the past 20 years can be bewildering. During the same period, the Institute of Laboratory Animal Resources (ILAR) has charted a responsible, balanced course by producing highly respected, science-based guidelines and advice for the biological research community. Many of ILAR's documents have won international acclaim. Yet the multiplication of regulations continues, coming from an alphabet soup of national or international organisations whose committees may sometimes decide to start afresh without the benefit of already tested knowledge.

Guidelines and regulations must retain the flexibility for improvement as knowledge advances. Therefore, it is vital to adopt performance based standards that are open to improvement and not engineering standards that can block or inhibit further progress. Researchers welcome appropriate regulation in all aspects of animal research including experimental procedures and management standards. It is important to seek some consensus in international research regulations in order to enhance: (1) the quality of the science; (2) the need for efficiency in costs of research; (3) the need for similar standards to govern the care and use of animals involved in international research protocols; (4) facilitation of the movement and exchange of research animals and animal products; (5) conservation of the time of scientists on the bench; (6) the achievement of the optimal care of traditional or non-traditional animal or cell stocks, and (7) the need to identify the key research questions that will lead to further improvements in animal care and use.

The current acceleration of globalisation also effects primate biology, the intellectual property related to research and the benefit sharing agreements that are essential between countries. As we witness such developments with the World Trade Organisation, the North American Free Trade Agreement and other global and regional alliances, science-based advice needs to be included early in the process in order to influence and assist the development of rational, efficient guidelines, regulations and legislation both nationally and internationally. The results of such teamwork in the management of biological resources in general and of primate resources in particular will be vital for the responsible research and development of human health, pharmaceuticals, aspects of trade and also for animal welfare and conservation. A summary of laboratory animal care policies and regulations from Canada, Japan, New Zealand, United Kingdom and the United States, together with a keynote essay on the impact of international free trade agreements on animal research is published in the ILAR Journal (Volume 37, 55–91, 1995). For further reference, see http://dels.nas.edu/ilar/

Conclusions

In this brief discussion I have itemised some of the significant factors that require consideration in an overall framework for the research, ethics and regulation of primate biology. In doing so it is perhaps important to stress practical outcomes. Many of the processes of globalisation can carry great benefits, but they also carry urgency and risk over the next 50 years if a responsible balance is not achieved. The rise of human populations from one billion in the mid 19th century to 2.5 billion in the mid 20th century to a likely 12 billion in the mid 21st century is not a matter that can be ignored and will require every possible effort for sustainable balance if the biota and the other integrated systems of the earth are not to fragment and become nonviable. Within this overall challenge, it is vital that research proceeds, and that our ethical and regulatory framework also advances. The role of primate biology in the overall challenge is significant, both in our understanding of fundamental factors affecting human and animal systems and health and also as a benchmark and key indicator in the development of culture and civilisation.

Acknowledgements

I thank my many friends and colleagues in primatology with whom debates and shared experience on these issues have shaped the views presented here. I acknowledge the formative and learning influences that I have enjoyed through my long association with Institutions and colleagues in Kenya and Uganda. I am grateful to Dr Paul Van Look, Director of the Department of Reproductive Health Research and of the Special Program of Research, Research Training and Development in Human Reproduction of the World Health Organisation for a research partnership over 25 years and for support for me at this conference.

Baboon and Other Nonhuman Primate Models for Research in Reproduction

4

Reproductive Behavior in Wild Baboons

S.C. Alberts[a,c], *J. Altmann*[b,c,d]

[a]Department of Biology, Duke University, Durham, NC, [b]Department of Ecology and Evolutionary Biology, Princeton University, Princeton, NJ, USA; [c]Institute for Primate Research, National Museums of Kenya, Nairobi, Kenya; [d]Department of Conservation Biology, Brookfield Zoo, Brookfield, IL, USA

The diversity and duration of field research on wild baboon populations, at sites from Ethiopia through South Africa, is unparalleled among mammals. Three sites have published analyses that draw on two decades or more of data [1–3], and two others have data from the same site over multiple years [4, 5]. The baboon population of Amboseli, Kenya, is the subject of particularly intensive and diverse long-term investigation (see http://www.princeton.edu/~baboon and bibliography therein). Because baboons are also among the most well-studied primates in the laboratory, the opportunity for comparative and synthetic work is enormous, but is just beginning to be realized.

The potential for significant synthesis is further enhanced by the fact that baboons share important evolved characteristics with humans. Like humans, baboons have adapted to a very wide range of environments, from near desert to temperate montane grasslands to moist evergreen forest; they have thereby achieved a nearly continental distribution in Africa [6]. Also like humans, but unlike the large majority of primate (and other mammal) species, baboons show virtually no seasonality in reproduction [7]. In other words, baboons, like humans, have both adapted to diverse habitats and have broken free of the seasonal constraints of these habitats in major aspects of their life histories. Finally, baboons, like humans, exhibit a highly flexible social system, which plays a key role in their adaptability [3].

Baboon Ecology

Baboons (members of the genus *Papio*) are increasingly treated as one species *(P. hamadryas)* with at least five subspecies *(hamadryas, anubis, cynocephalus, ursinus* and *papio)*, each of which has a distinct geographical range and some unique morphological features [6]. These large, diurnal, semi-terrestrial monkeys are highly flexible foragers. They are eclectic and omnivorous feeders, but this omnivory is combined with great discrimination. They feed very selectively, often choosing a small component of a plant and forgoing the remainder, or focusing on a single species within a genus. Plants are the most important source of nutrients; invertebrate and vertebrate animals are eaten but contribute relatively little in calories and protein. In the habitats in which baboons are best studied (savannahs), they rely heavily on grasses, consuming both the underground storage organs (corms) and the leaves [8–11].

Savannah baboons (the common term for all subspecies except *P. h. hamadryas*) live in stable social groups of 20–100 members, including multiple adults and juveniles of both sexes. Hamadryas baboons *(P. h. hamadryas)* in the horn of Africa deviate markedly from this social pattern. Their basic social unit consists of (usually) a single male and several females and dependent young, but these 'one-male units' coalesce at the end of each day into hierarchically structured larger social units. The discussion here will focus on savannah baboons, but for more information on hamadryas baboons [12, 13].

The larger African carnivores – leopards, lions, and hyenas – prey on baboons and pose particularly high risk at night. In each habitat where they are found, members of a baboon group sleep close together either on cliff edges or high in those trees in their habitat that would be the most difficult for a predator to climb. For baboons, patterns of group movement and activity are determined by the distribution of sparsely scattered nighttime sleeping sites, of potable water and food resources, and of predators. For instance, in Amboseli, Kenya, baboons awaken and descend from their sleeping trees shortly after dawn. For the next 11–12 h, they spend almost 75% of their time foraging – feeding or traveling to food – across their short-grassland savannah habitat, approximately 10% socializing, and the remainder resting, often in a mid-day 'siesta' [3].

Baboon Social Behavior

Savannah baboons exhibit two very striking social characteristics, which they share with a number of other mammals. First, they exhibit a matrilineal social structure [7]. Older females, their mature daughters, and their immature offspring of both sexes live in close proximity, and maternal relatedness is a major predictor of social interactions within groups. In study after study of matrilineal societies, the strong predictive power of maternal relatedness for social associations has been demonstrated; other examples of species that exhibit this type of social organization include elephants [14], spotted hyenas [15], a number of cetaceans [16], ground squirrels [17], and red deer [18]. Although details of social structure differ across taxa, matrilineal relationships in baboons predict dominance rank, support in agonistic (aggressive-submissive) coalitions, proximity during foraging and other activities, and grooming relationships [7, 19, 20]. Because of the ubiquity of strong kin bonds in these societies, researchers have inferred that social relationships with kin are important for successful reproduction and survival, and empirical evidence is mounting to support this inference.

Second, baboons exhibit strong linear dominance hierarchies in both sexes, in which relationships are typically transitive and circularities are rare [7, 19, 20]. That is, for any pair of animals in a social group, the outcomes of agonistic interactions between them are relatively predictable and stable over periods of months (in the case of adult males) or years (in the case of adult females). Further, if animal A wins disputes with animal B and B wins disputes with C, one can generally predict that A will win disputes with C. Among female baboons (and other cercopithecines), daughters typically attain adult rank just below their mothers, and the relative ranking of families tends to be stable, even across several generations. Among males, dominance rank is independent of maternal rank, and changes markedly with age; males reach their highest rank between 8 and 10 years of age (in their prime), and fall in rank steadily after this age [2, 21]. For both sexes, dominance rank predicts competitive ability in at least some contexts (see below).

Reproductive Behavior and Life History

Wild female baboons usually reach menarche early in their fifth year (roughly a year after attaining their dominance rank), and during adulthood they have highly visible sexual swellings that indicate ovarian cycle phase (follicular, luteal, menstrual), as well as skin-color changes that indicate pregnancy. They produce a single offspring approximately every 2 years from age 6 through at least age 20, if they live that long [3].

In all wild populations examined, females may conceive and give birth in any month [22–24]. However, births in some populations exhibit slight seasonal peaks. For instance, in Amboseli, Kenya, births occur more often than expected during the dry season: 49% (242/495) of live births in this population between 1971 and 2000 occurred in the five months of the long dry season, June through October, compared to 42% (206) expected (G test of goodness of fit, $G = 10.64$, $p < 0.005$). This corresponds to conceptions occurring most often from February through May, during the typical months of the 'long rainy season'. However, even this weak birth clustering seen in Amboseli may reflect seasonal and annual variability in risks and resources that are important to survival and reproduction. Birth is a milestone event and often the only available measure for reproductive seasonality. Nonetheless, it reflects earlier milestone dates and constrained life stages, ones that may be more immediately dependent on variability in risks and resources. Because gestation duration is the least variable life history stage, dates of live births are almost entirely determined by conception dates. Conception dates, in turn, are determined by the timing of the onset of cycling (menarche or post-partum) and the probability of conception. In baboons, these are both highly variable as a result of both stochastic and deterministic processes, including those influenced by resources and risks.

Males differ from females in several life history traits. Baboons exhibit bimaturism – males do not become fully adult until about 8–10 years, 3–5 years after their female peers [25]. Related to their bimaturism is their sexual dimorphism, with males nearly twice the body mass of adult females. This dimorphism is achieved by a male growth spurt that occurs largely after both sexes reach puberty and after females start reproducing and stop growing [26, 27]. Further, while females remain throughout their lives in their group of birth, males disperse and move into other social groups [28, 29]. This male dispersal generally occurs first at around 8 years of age and then repeatedly over the course of their lives (about every two years on average in the Amboseli population [29]). Male dispersal patterns appear to be strongly influenced by the costs and opportunities associated with remaining in one group versus moving to another. These costs and opportunities include the risk of mortality during the dispersal episode, the availability of sexually cycling females, and the risk of inbreeding that a male experiences if he remains in his group of birth [28, 29].

Savannah baboons mate in the context of mate-guarding episodes, generally known as sexual consortships, which are conspicuous episodes of close, persistent following of females by males accompanied by sexual activity [30–32]. Mate-guarding episodes may last from several hours to several days, and typically occur while females are in the second half of the follicular phase of the sexual cycle. However, relationships between adult males and adult females are not limited to the follicular phase of the female sexual cycle; in many wild populations, persistent affiliative relationships, which may last for months or even years, have been documented between males and females [24, 33].

Dominance and Male Reproductive Behavior

Dominance rank has a major impact on the mating activity of male baboons: high ranking males experience higher mating success [21, 30–32]. In general, the dominance hierarchy functions as a queue in which males wait for mating opportunities. That is, when only one female in a social group is sexually receptive, only the highest ranking male will mate with her; when two females are simultaneously receptive, the highest and second ranking males will mate, and so on. However, this queuing system breaks down under some circumstances, with the consequence that lower ranking males are sometimes able to garner considerable access to sexually receptive females. This happens when lower-ranking males form frequent male-male coalitions to team up against high-ranking males, when the highest-ranking male has only recently attained high status, and/or when social groups are very large. Male-male coalitions in particular appear to constitute a major 'alternative mating strategy' for male baboons [21, 31, 32]. However, male-female friendships may also play an important role in male mating success. In particular, lower-ranking males may experience enhanced mating success if females express strong preferences for them [33].

Dominance and Female Reproductive Behavior

Dominance rank also affects female reproductive performance, although its effects are less dramatic than for males [19]. For instance, in Amboseli, higher ranking females both reach menarche earlier and experience higher reproductive rates [3]. These rank effects may be mediated by differences in offspring growth rate; offspring of high-ranking females in Amboseli grew more rapidly, and rapid growth predicted early maturity. A 10-rank difference in dominance status was associated with a 0.2-year (2.5-month) difference in age of menarche and a 0.13-year (1.5-month) difference in interbirth interval. Such rank-associated differences in reproductive performance have been reported in several wild baboon populations. Thus, through attainment and maintenance of high status, baboon females accrue fitness benefits from both enhanced fertility and offspring quality. Although relative status of females within a group is ordinarily transmitted with great fidelity between generations, a low-ranking female can reduce the number of others who dominate her or her daughters by participating in a group fission, and can thereby escape the costs of low status [3].

Until very recently the causes and consequences of reproductive behavior in wild baboon populations were studied almost exclusively through the use of observational techniques so as not to disrupt the social systems under investigation. Now, however, dramatic new ground is being broken by completely non-invasive studies of genetic relatedness (including paternity determination) [34–36] and of reproductive physiology [37, 38]. These non-invasive studies depend on newly-developed techniques for extracting both DNA and steroid metabolites from freshly deposited fecal samples from known individuals. These breakthroughs will allow us to refine even further our understanding of baboon behavior and bring closer a true synthesis of field and laboratory studies.

References

1 Rhine RJ, Norton GW, Wasser SK: Lifetime reproductive success, longevity, and reproductive life history of female yellow baboons *(Papio cynocephalus)* of Mikumi National Park, Tanzania. Amer J Primatol 2000;51: 229–242.

2 Packer C, Collins DA, Eberly LE: Problems with primate sex ratios. Phil Trans R Soc 2000;355:1627–1635.

3 Altmann J, Alberts SC: Intraspecific variability in fertility and offspring survival in a nonhuman primate: behavioral control of ecological and social sources; in Wachter K, Bulatao R: Offspring: The Biodemography of Fertility and Family Behavior. Washington, National Academy Press, 2003, pp 140–169.

4 Bercovitch FB, Strum SC: Dominance rank, resource availability, and reproductive maturation in female savanna baboons. Behav Ecol Sociobiol 1993;33:313–318.

5 Bulger J, Hamilton WJ, III: Rank and density correlates of inclusive fitness measures in a natural chacma baboon *(Papio ursinus)* troop. Int J Primatol 1987;8:635–650.

6 Jolly CJ: Species, subspecies, and baboon systematics; in Kimbel, WH, Martin, LB: Species, Species Concepts, and Primate Evolution. New York, Plenum Press, 1993, pp 67–107.

7 Melnick DJ, Pearl MC: Cercopithecines in multimale groups: genetic diversity and population structure; in Smuts BB, Cheney DL, Seyfarth R, Wrangham RW, Struhsaker TT: Primate Societies. Chicago, University of Chicago Press, 1987, pp 121–134.

8 Post DG: Feeding behavior of yellow baboons *(Papio cynocephalus)* in the Amboseli National Park, Kenya. Int J Primatol 1982;3:403–430.

9 Norton GW, Rhine RJ, Wynn GW, Wynn RD: Baboon diet: a five-year study of stability and variability in the plant feeding and habitat of the yellow baboons *(Papio cynocephalus)* of Mikumi National Park, Tanzania. Folia Primatol 1987;48:78–120.

10 Altmann SA: Foraging for Survival. Chicago: University of Chicago Press, 1998.

11 Byrne RW, Whiten A, Henzi SP, McCulloch FM: Nutritional constraints on mountain baboons *(Papio ursinus)*: implications for baboon socioecology. Behav Ecol Sociobiol 1993;33:233–246.

12 Stammbach E: Desert, forest and montane baboons: multilevel societies; in Smuts, BB, Cheney DL, Seyfarth R, Wrangham RW, Struhsaker TT: Primate Societies. Chicago, University of Chicago Press, 1987, pp 112–120.

13 Kummer H: Social Organization of Hamadryas Baboons. Chicago, University of Chicago Press, 1968.

14 Moss CJ, Poole JH: Relationships and social structure of African elephants; in Hinde RA: Primate Social Relationships. Sunderland, Sinauer, 1983, pp 315–325.

15 Engh AL, Esch K, Smale L, Holekamp KE: Mechanisms of maternal rank 'inheritance' in the spotted hyaena, *Crocuta crocuta*. Anim Behav 2000;60: 323–332.

16 Bigg MA, Olesiuk PF, Ellis GM, Ford JKB, Balcomb KC: Social organization and genealogy of resident killer whales *(Orcinus orca)* in the coastal waters of British Columbia and Washington State; in Hammond PS, Mizroch SA, Donovan GP: Individual Recognition of Cetaceans: Use of Photo-identification and Other Techniques to Estimate Population Parameters. Cambridge, International Whaling Commission, 1990, pp 383–405.

17 Sherman PW: Kinship, demography, and Belding's ground squirrel nepotism. Behav Ecol Sociobiol 1981;8:251–259.

18 Clutton-Brock TH, Guinness FE, Albon SD: Red Deer: Behavior and Ecology of Two Sexes. Chicago, University of Chicago Press, 1982.

19 Silk JB: Social behavior in evolutionary perspective; in Smuts BB, Cheney DL, Seyfarth RM, Wrangham RW, Struhsaker TT: Primate Societies. Chicago, University of Chicago Press, 1987, pp 318–329.

20 Walters JR, Seyfarth RM: Conflict and cooperation; in Smuts BB, Cheney DL, Seyfarth R, Wrangham RW, Struhsaker TT: Primate Societies. Chicago, University of Chicago Press, 1987, pp 306–317.

21 Alberts SC, Watts HE, Altmann J: Queuing and queue-jumping: long-term patterns of reproductive skew in male baboons. Anim Behav, in press.

22 Bercovitch FB, Harding RSO: Annual birth patterns of savanna baboons *(Papio cynocephalus anubis)* over a ten-year period at Gilgil, Kenya. Folia Primatol 1993;61:115–122.

23 Bentley-Condit VK, Smith EO: Female reproductive parameters of Tana River yellow baboons. Int J Primatol 1997;18:581–596.

24 Altmann J: Baboon Mothers and Infants. Cambridge, Harvard University Press, 1980.

25 Alberts SC, Altmann J: Preparation and activation: determinants of age at reproductive maturity in male baboons. Behav Ecol Sociobiol 1995;36: 397–406.

26 Strum SC: Weight and age in wild olive baboons. Amer J Primatol 1991;25: 219–237.

27 Altmann J, Alberts S: Body mass and growth rates in a wild primate population. Oecologia 1987;72:15–20.

28 Packer C: Inter-troop transfer and inbreeding avoidance in *Papio anubis*. Anim Behav 1979;27:1–36.

29 Alberts SC, Altmann J: Balancing costs and opportunities: dispersal in male baboons. Amer Natur 1995;145:279–306.

30 Packer C: Male dominance and reproductive activity in *Papio anubis*. Anim Behav 1979;27:37–45.
31 Bulger JB: Dominance rank and access to estrous females in male savanna baboons. Behaviour 1993;127:67–103.
32 Hausfater G: Dominance and Reproduction in Baboons *(Papio cynocephalus)*. Basel, Karger, 1975.
33 Smuts BB: Sex and Friendship in Baboons. Hawthorn, Aldine, 1985.
34 Altmann J, Alberts SC, Haines SA, Dubach J, Muruthi P, Coote T, et al: Behavior predicts genetic structure in a wild primate group. Proc Natl Acad Sci USA 1996;93:5797–5801.
35 Smith KL, Alberts SC, Bayes MK, Bruford MW, Altmann J, Ober C: Cross-species amplification, non-invasive genotyping, and non-Medelian inheritance of human STRPs in savannah baboons. Amer J Primatol 2000;51: 219–227.
36 Smith K, Alberts SC, Altmann J: Wild female baboons bias their social behaviour towards paternal half-sisters. Proc R Soc Lond B Biol Sci 2003; 270:503–510.
37 Khan MZ, Altmann J, Isani SS, Yu J: A matter of time: evaluating the storage of fecal samples for steroid analysis. Gen Comp Endocrinol 2002; 128:57–64.
38 Wasser SK, Hunt KE, Brown JL, Cooper K, Crockett CM, Bechert U, et al: A generalized fecal glucocorticoid assay for use in a diverse array of non-domestic mammalian and avian species. Gen Comp Endocrinol 2000;120: 260–275.

5

Use of a Baboon Model for Research in Human Contraception

V.C. Stevens

Department of Obstetrics and Gynecology,
Ohio State University, Columbus, OH, USA

Research and development efforts to provide new or improved methods for human contraception must employ preclinical studies demonstrating both efficacy and safety of the proposed product. Depending on the nature of the compound(s) or devices involved, various animal models can be used to provide the necessary data to satisfy drug regulatory authority requirements. In certain situations, rodents or other common laboratory animals will suffice to meet these needs, but in many cases, it is necessary to employ a non-human primate that closely resembles humans in their anatomy and reproductive physiology. In such cases, several factors must be considered in selecting the most suitable non-human primate species. First, the animals must not be in danger of extinction, and can be acquired in relatively large numbers at an affordable cost. Second, these must be adaptable to captivity under conditions that will not effect their survival or reproductive function. Third, the animals must be of a size and weight that will allow their physical manipulation without excessive equipment or numbers of staff. Fourth and finally, the species must be sufficiently abundant that sacrifices of experimental animals for tissue pathological examinations will not violate ethical or financial bounds.

Baboons, various species of the genus *Papio,* are still abundant in Africa and while expensive, can be obtained commercially in Europe, Asia, and North America. They adapt well to captivity and under ideal conditions, can be somewhat domesticated. Experienced handlers equipped with appropriate cages and protective clothing can easily conduct blood samplings and treatments. Normally, wild-caught animals will exhibit adaptive behavior and physiology within 3–6 months of captivity. A general indicator of adaptation for female baboons is their demonstration of regular, predictable fluctuations of sex skin turgescence and deturgescence during their menstrual cycles. While male baboon adaptation is more difficult to assess, they are considered adapted when they readily exhibit active breeding behavior.

Table 1. Some parameters observed during 700 baboon menstrual cycles

Parameter	Mean	s.d.
Cycle length, days	33.4	2.1
Menses, days	3.2	1.0
Menses to maximum sex skin turgescence, days	6.2	1.7
Maximum sex skin turgescence, days	12.1	2.4
Sex skin turgescence to menstruation, days	14.6	1.6
Estradiol peak level to sex skin deturgescence, days	3.3	1.2
LH peak level to sex skin deturgescence, days	2.3	0.8
LH peak level to menstruation, days	16.9	1.1
Progesterone peak level to menstruation, days	9.0	1.3
Maximum level estradiol per cycle, pg/ml	245.0	30.5
Maximum level progesterone per cycle, ng/ml	11.5	2.3

When testing potential new contraceptives that effect ovulation, other menstrual cycle events or endocrine parameters of the hypothalamic, pituitary, gonadal axis, the female baboon exhibits characteristics very similar to women. As can be seen in table 1, several events are quite predictable in a population of well-stabilized captive females. While the follicular phase and overall cycle length is slightly longer in baboons than in women, the menstrual cycles in both are otherwise nearly identical. The only major difference is the lack of a luteal phase elevation of estrogens in baboons typically found in women.

During pregnancy endocrine patterns in baboons are also very similar to humans excepting that the principal estrogen produced during gestation in the baboon is estrone, not estriol, as found in human pregnancies. When an adjustment is made for the shortened gestation in baboons (about 6 months) compared to humans, the patterns of gestational hormones mimic those in women. The main advantage of using baboons for contraceptive method assessment is their ability to breed in captivity at a high rate. This rate is related to the level of colony management and culling of infertile animals from the breeding population (table 2). A conception rate of 50–80% per each exposure to pregnancy in a well-managed colony exceeds that in young, healthy women in normal marital relationships. This high rate permits a statistically valid anti-fertility trial with relatively small numbers of animals in treatment and control groups.

Numerous studies of effects of anti-fertility compounds and devices have been conducted during the past thirty-five years in captive baboon colonies. Female baboons fitted with copper-coated intrauterine devices showed no changes in liver functions or cellular ultrastructure, as was once found in an unreported human case, thus relieving concern over its continued use for human contraception [1]. The potential use of a gonadotropin-releasing agonist as an anti-fertility compound in non-pregnant females, and as an abortifacient during pregnancy was also tested [2]. Another study tested the anti-fertility effects of vaginal applications of immunoglobulins [3]. There have been a variety of other studies reported in which baboons were

Table 2. Pregnancy rates in female baboons in a well-managed colony

Menstrual cycle characterization	Number of baboons[a]	Number of matings	Number of pregnancies	Fertility rate, %
Optimal[b]	187	868	682	78.6
Suboptimal[c]	81	268	188	70.1
Minimal[d]	47	220	115	52.3

[a] A total of 12 male baboons of varying ages were used between 1966 and 1994. No male was used in subsequent matings if he failed to fertilize a female in his last three attempts.
[b] Female baboons monitored for >20 menstrual cycles before mating. Animals without cycle regularity (+ more than 3 days in cycle length) or with more than one anovulatory cycle or non-patent oviducts were culled.
[c] Female baboons monitored for >6 menstrual cycles before mating. Animals with any anovulatory cycle or variation in cycle length for +4 days were culled from the group.
[d] Female baboons monitored for at least three menstrual cycles before mating. All cycles must have been ovulatory.

used for preclinical testing of experimental contraceptive compounds, but the major use of these animals in the author's laboratory has been in the development and testing of anti-fertility vaccines for use by women [4].

The testing of anti-fertility vaccines, intended for use by women, requires an animal model with significant similarities to humans. Not only must their reproductive characteristics be similar, but also their immune system and reproductive proteins must be close in structure and function. Antibodies generated against human reproductive antigens, with a few exceptions, will not react with those of rodents, rabbits, or other commonly used laboratory animals.

Immune recognition of protein hormones or gonadal antigens are often highly species specific and antibodies to human antigens infrequently cross-react with analogous components in phylogenetically lower animals. Thus, a human vaccine antigen can be tested in the model species only if there is sufficient cross-reactivity between the analogous molecules in both species. Alternatively, two vaccines must be developed, one specific to the human antigen, and one specific to the animal model antigen. This scenario is very expensive and time-consuming and difficult to justify to funding sources.

Studies in the author's laboratory, dating some twenty-eight years ago, showed that antibodies against the beta subunit of human chorionic gonadotropin (hCG) were capable of disrupting normal pregnancy in baboons, and that females immunized against the subunit were incapable of sustaining pregnancy [5]. These early studies showed the potential of active immunization of women as a new approach to fertility regulation. The problem found with immunization against hCG beta subunit was that the antibodies raised against the hormone subunit reacted not only with hCG, but also with the human pituitary hormone, luteinizing hormone (hLH). Fearing that immunization of women with this molecule might disrupt the normal menstrual cycle, or even worse, cause autoimmune damage to the pituitary, further research and development of a vaccine using the hCG beta subunit was discontinued in our laboratory. However, when the amino acid sequences of glycoprotein hormone subunits were described, it was revealed that a portion of the hCG beta subunit contained a peptide sequence not found in the hLH beta subunit. Since that time, extensive studies have been conducted to develop a safe and effective vaccine for human use, using hCG specific peptide antigens. Baboons have been invaluable as a model species in these studies

Several years of study were necessary to synthesize peptides, formulate an acceptable immunogen, and test the immunogencity and hormone neutralizing capacity of antibodies to a large number of vaccine components and formulations. The results of these studies showed that antibodies raised to synthetic peptides representing the carboxyl terminal region of the hCG beta subunit react well with intact hCG, but not at all with hLH or other pituitary hormones. Further, it was found that peptides of this region must have sufficient length to include at least two immunological epitopes for generating antibodies with more than one binding site in order to obtain hCG neutralizing effects. As the carboxyl end of hCG beta subunit is not involved in interactions with hCG gonadal receptors, clearance from the circulation by multiple immunoglobulin binding is necessary to neutralize hormonal functions. Initially, an experimental vaccine formulation consisting of a synthetic peptide representing amino acids 109–145 of hCG beta subunit conjugated to tetanus toxoid was tested in female baboons. While the vaccine employed Freund's Complete adjuvant and was not acceptable for human application, high levels of antibodies reactive with hCG were induced by it [6], and several observations important for further vaccine development were made. While none of the baboons' generated antibodies reacted with hLH or baboon LH, a low level of antibodies reactive with baboon CG was found (about 5% as much as to hCG). Despite this low antibody level, a marked reduction in fertility was observed. Passive immunization of non-immunized pregnant baboons with antisera from immunized ones resulted in immediate abortions. This clearly indicated that antibodies to the hCG peptide neutralized baboon CG in early pregnancy, as well as antisera to the entire beta subunit. A further observation was that there were significant delayed hypersensitivity (DTH) reactions to the carrier protein tetanus toxoid.

Collectively, these observations provided information for the design of a vaccine formulation that could potentially be tested in women. Subsequently, a vaccine was developed using the same peptide sequence, but employing a different carrier protein (diphtheria toxoid) and using an adjuvant formulation less reactive at the injection site than Freund's Compete adjuvant. Another anti-fertility trial using these latter formulations was tested in 15 well-stabilized female baboons. The encouraging data from this trial stimulated further studies of the immunosafety of this vaccine formulation in some 50 baboons, including extensive histopathological examinations, and a Phase I clinical trial [7]. The clinical study revealed that the vaccine was safe and elicited antibodies in all subjects, but the duration of elevated antibodies was insufficient for application to birth control. Since that time, several studies have been conducted to develop improved immunogens and new vaccine delivery systems, which will induce long lasting levels of antibodies from a single injection without unacceptable tissue reactions at the injection site [8]. Baboons have been employed in many of these studies.

A number of other experimental anti-fertility vaccines besides the hCG vaccines have been tested in baboons. These include one against the sperm enzyme $LDHC_4$ [9, 10], a sperm surface antigen [11] and a zona pellucida antigen [12]. While none of these vaccines, excepting hCG vaccines has been tested in humans, their evaluation in baboons has given significant insight to their efficacy, safety, and problems needed to be resolved before application to human use.

Acknowledgement

Studies on hCG vaccine development received financial support from the Special Programme of Research, Development, and Research Training in Human Reproduction, World Health Organization.

References

1 Hagenfeldt K, Shelden RM, Stevens V: Absence of Copper IUD on Baboon Menstrual Cycle Parameters, Liver Enzymes or Liver Cellular Structure. Unpublished report to World Health Organization 1978.
2 Vickery BH, McRae GI, and Stevens VC: Suppression of luteal and placental function in pregnant baboons with agonist analogs of luteinizing hormone releasing hormones. Fertil Steril 1981;36:664–668.
3 Beck LR, Boots L, and Stevens VC: Absorption of antibodies from the baboon vagina. Biol Reprod 1975;13:10–15.
4 Stevens VC: Some reproductive studies in the baboon. Human Reproduction Update 1997;3:533–540.
5 Stevens VC: Fertility control through active immunization using placental proteins. Acta Endocrinol 1975;78:357–375.
6 Stevens VC, Powell JE, Lee AC, and Griffin PD: Antifertility effects from immunization of female baboons with C-terminal peptides of hCG beta subunit. Fertil Steril 1981;36:98–105.
7 Jones WR, Braley J, Judd SF, et al: Phase I clinical trial of a World Health Organization birth control vaccine. Lancet 1988;i:1295–1298.
8 Cui C, Stevens VC, Schewendeman SP: Immunogencity of Synthetic Human Peptide Antigen fronm Encapsulated and Surface- Conjugated Poly(D,L-lactide-co-glycolide) Microspheres Proceed Int Symp Control Rel Bioact Mater 2002;29:674–675.
9 Goldberg E, Wheat TE, Powell JE, and Stevens VC: Reduction of fertility in female baboons immunized with lactate dehydrogenase-C4. Fertil Steril 1980;35:214–217.
10 O'Hern PA, Bombra CS, Ishakia M, and Goldberg E: Reversible contraception in female baboons immunized with a synthetic epitope of sperm-specific lactate dehydrogenase. Biol Reprod 1995;52:331–334.
11 Herr JC, Wright RM, Klotz K, et al: Immunogenicity of SP-10 fusion proteins in female baboons. In Talwar, GP, Rao, KVS. and Chuhan VS (eds): Proceedings of Symposium on Recombinant and Synthetic vaccines Narosa Publishing House, New Delhi 1994, pp 239–251.
12 Dunbar BS, Powell JE, and Stevens VC: Use of a synthetic peptide adjuvant for the immunization of baboons with denatured and deglycosylated pig zona pellucida glycoproteins. Fertil Steril 1989;52:311–318.

6
Subhuman Primates as Models for the Development of Male Contraceptives

G. van der Horst[a], J. Seier[b], M.C. Mdhluli[b]

[a]Department of Medical Biosciences, University of the Western Cape, Bellville, and [b]Primate Unit, Medical Research Council of South Africa, Parow Valley, South Africa

Introduction and Review

The main aims of this paper are twofold. Firstly, it presents a review of some of the advances made in using subhuman primates as models for human male contraception. Secondly, it concentrates on the suitability of and advances made in using the African green monkey, *Chlorocebus aethiops* (until recently known as *Cercopithecus aethiops*), also known as the vervet monkey, in contraceptive studies.

Table 1. Some subhuman primate species used for various contraceptive approaches

Contraceptive approach	Subhuman primates used	References
Hormonal		
17-beta estradiol and levonorgestrel	Cynomolgus monkey	1
7-alpha-methyl 19 nortestosterone (MENT)	Bonnet monkey	2
GnRH	Marmoset	3
Immunological		
Immunogen to CD 52 protein	Chimpanzee	4
Immunogen to LDH-4(C)	Baboon	5
Riboflavin Carrier Protein (RCP) vaccine	Bonnet monkey	6
Various immunogens to acrosome	Rhesus monkey	7
Immunogen to PH-20 protein	Cynomolgus and	8
	Macaque	9
Immunogen to ZP3	Bonnet monkey	
Mechanical		
Vasectomy	Chimpanzee	10
Vas occlusion with SMA	Langurs	11
Other/natural plant products isolated		
Pentacyclic triterpene, oleanolic acid	African green monkey	Current study

Hormonal, immunological, mechanical and other diverse approaches have been used to test various subhuman primate models for their suitability as models for human male contraception. Table 1 lists some of the subhuman primate species used and the contraceptive approach utilized.

Hormonal approaches including various reproductive steroids [1], synthetic steroids [2], and GnRH analogues [3] have been used with various levels of success as male contraceptives. Synthetic steroids such as MENT has the advantage that no additional testosterone therapy is required. Oral contraceptive studies in subhuman primates suggest similarities with humans, however, it is not clear what the long-term effects of exogenous androgens may be particularly in terms of the risk of prostate cancer.

The development of immunological approaches in male contraception has increased during the last decade particularly in view of the improvement in the technology of molecular biology in general. While this approach allows the investigator to be very specific in targeting a particular male reproductive protein such as LDH-4(C) (where immunogen is specific in mice, baboon and human) it has several disadvantages. According to McCauley et al. [4], sperm agglutination antigen-1 (SAGA-1) is a human male reproductive tract glycoform of CD52 which in a modified form (CD52 N-linked oligosaccharide chains – forming carbohydrate epitope) is localized over the entire surface of human spermatozoa in the testis and epididymis. Collectively, sperm surface localization, antibody inhibition of sperm function, and potential reproductive- tissue specificity identify SAGA-1 as an attractive candidate contraceptive immunogen [4]. However, the immunogen developed is specific to compromise fertility in the chimpanzee and human but not in bonnet monkeys, macaques and the baboon [4]. In this example of immunological contraception, most subhuman primates do not represent good models for human male contraception. Furthermore, in some instances cross-reaction of the immunogen with other cells/tissues in the body represents problems [12]. Vasectomy is not reversible in all humans but vas occlusion with styrene maleic acid in langurs appears to be fully reversible [11].

Despite some of the problems outlined above both hormonal as well as immunological approaches show potential in male contraception and in many instances several subhuman primate species appear to be appropriate models to study.

The African green monkey is the one of the most widely distributed non-human primates in the world of all African Monkeys [13]. They occur from southern Ethiopia and Somalia to the Cape region of South Africa [14]. It is also one of very few subhuman primate species that is successfully bred under controlled captive conditions in the Primate Unit of the MRC, Parow Valley, South Africa. In this study we will show the potential of the African green monkey as an appropriate model for human male contraceptive studies and present some data on progress that has been made with a potential anti-fertility substance, oleanolic acid, isolated from cloves.

Materials and Methods

Structural Studies. Ten reproductive systems (testes to penis) of the African green monkey were used for routine histological studies and scanning electron microscopy and the detailed structure of each part of the reproductive system was compared to that of humans.

Sperm Maturation. Sperm were isolated from the caput, corpus and cauda epididymidis as well as from the vas deferens and the ejaculate, suspended in Ham's F10 and sperm motility was studied quantitatively at 37 °C. Sperm morphology was also studied as a function of sperm maturation in the epididymis.

Contraceptive Studies. Three different doses of oleanolic acid (OA) (4–25 mg/kg) were orally given on a daily basis for six months to African green monkeys (three per group) and three served as controls. Semen parameters, acrosome status (fluorescence microscopy), behaviour and clinical chemistry (major heart, liver, kidney enzymes as well as haematological parameters) were studied every two weeks over the study period. Mating experiments were performed using all experimental and control monkeys over two menstrual cycles at the end of the study.

Results and Discussion

Structural. The histological structure of the African green monkey reproductive system does not differ in any major respect from that of humans.

Sperm Maturation. The percentage motile sperm in culture medium of sperm isolated from different parts of the reproductive system of the African green monkey show remarkable agreement with published literature on several other subhuman primate species as well as with the human [15, 16]. The major sperm morphological changes such as the presence and movement of the cytoplasmic droplet and acrosomal changes during epididymal transit resemble human sperm maturation.

Contraceptive Studies. Most semen parameters were not affected by OA administration except for acrosomal status. OA administration to rats indicated that acrosomal structure as studied by means of scanning electron microscopy was severely affected [17] and implies a similar mechanism of action in African green monkeys.

Mating experiments yielded offspring in the control group but none in the OA treated group receiving the lowest dose. However, offspring were born in groups receiving higher doses of OA. While it is clear that OA has anti-fertility properties some of the above results are contradictory probably due to the small sample sizes that were employed. The preliminary results may suggest that OA has a biphasic effect that is dose dependant. There were also no changes in the behaviour of the monkeys. These results are in agreement with mating studies in the rat which also suggested that libido was not affected [17].

Clinical chemistry analysis indicated that there were no significant changes in virtually all parameters measured between control and experimental monkeys. In addition the clinical chemistry parameters of the African green monkeys and those of humans are almost similar and show furthermore that the African green monkey represent a good model in this context.

Conclusion

Based on detailed structural studies of the male reproductive system, sperm maturation, clinical chemistry parameters as well as haematology, the African green monkey appears to be a good model to use in human male contraceptive studies.

Preliminary studies furthermore show that the pentacyclic triterpene, oleanolic acid, show potential to be used as a human male contraceptive, particularly in view that OA does not appear to be toxic or have any side effects on the African green monkeys.

References

1 Adams MR, Clarkson TB, Koritnik DR, Nash HA: Contraceptive steroids and coronary artery atherosclerosis in cynomolgus macaques. Fertil Steril 1987;47:1010–1018.
2 Ramachandra SG, Ramesh V, Krishnamurthy HN, Kumar N, Sundaram K, Hardy MP, Rao AJ: Effect of chronic administration of 7alpha-methyl-19-nortestosterone on serum testosterone, number of spermatozoa and fertility in adult male bonnet monkeys *(Macaca radiata)*. Reproduction 2002; 124:301–309.
3 Fraser HM, Recio R, Conn PM, Lunn SF: Gonadotropin-releasing hormone antagonist for postpartum contraception: outcome for the mother and male offspring in the marmoset. J Clin Endocrinol Metab 1994;78: 121–125.
4 McCauley TC, Kurth BE, Norton EJ, Klotz KL, Westbrook VA, Rao AJ, Herr JC, Diekman AB: Analysis of a human sperm CD52 glycoform in primates: identification of an animal model for immunocontraceptive vaccine development. Biol Reprod 2002;66:1681–1688.
5 Goldberg E, VandeBerg JL, Mahony MC, Doncel GF: Immune response of male baboons to testis-specific LDH-C(4). Contraception 2001;64:93–98.
6 Adiga PR, Subramanian S, Rao J, Kumar M: Prospects of riboflavin carrier protein (RCP) as an antifertility vaccine in male and female mammals. Human Reprod Update 1997;3:325–334.
7 Archibong AE, Lee CY, Wolf DP: Functional characterization of the primate sperm acrosomal antigen (PSA-63). J Androl 1995;16:318–326.
8 Deng X, Meyers SA, Tollner TL, Yudin AI, Primakoff PD, He DN, Overstreet, JW: Immunological response of female macaques to the PH-20 sperm protein following injection of recombinant proteins or synthesized peptides. J Reprod Immunol 2002;54:93–115.
9 Afzalpurkar A, Shibahara H, Hasegawa A, Koyama K, Gupta SK: Immunoreactivity and in-vitro effect on human sperm-egg binding of antibodies against peptides corresponding to bonnet monkey zona pellucida-3 glycoprotein. Hum Reprod 1997;12:2664–2670.
10 Hoffman K, Howell S, Schwandt M, Fritz J: Vasectomy as a birth control modality for captive chimpanzees. Lab Animals (NY) 2002;31:45–48.
11 Lohiya NK, Manivannan B, Mishra PK, Pathak N, Balasubramanian SP: Intravasal contraception with styrene maleic anhydride and its noninvasive reversal in langur monkeys (Presbytis entellus entellus). Contraception 1998;58:119–128.
12 Gupta SK, Koothan PT: Relevance of immuno-contraceptive vaccines for population control. I. Hormonal immunocontraception. Arch Immunol Ther Exp (Warsz) 1990;38:47–60.
13 Eley RM: Reproductive biology of the vervet monkey *(Cercopithecus aethiops)*: Occasional papers of the national museums of Kenya, Utafiti 1992;4:1–33.

The Baboon as a Nonhuman Primate Model for the Study of Human Reproduction

14 Skinner JD, Smithers RHN: The Mammals of the southern African Subregion. University of Pretoria Press 1990; pp 159–164.
15 Yeung CH, Morrell JM, Cooper TG, Weinbauer GF, Hodges JK, Nieschlag E: Maturation of sperm motility in the epididymis of the common marmoset *(Callithrix jacchus)* and the cynomolgus monkey *(Macaca fascicularis)*. International J Androl 1996;19:113–121.
16 Yeung CH, Cooper TG, Oberpenning F, Schulze H, Nieschlag E: Changes in movement characteristics of human spermatozoa along the length of the epididymis. Biol Reprod 1993;49:274–280.
17 Mdhluli MC, van der Horst G: The effect of oleanolic acid on sperm motion characteristics and fertility of male Wistar rats. Laboratory Animals 2002;36:432–437.

7

The Male Vervet Monkey: Sperm Characteristics and Use in Reproductive Research

M.C. Mdhluli[a], J.V. Seier[a], G. van der Horst[b]

[a]Primate Unit, Diabetes Research Group, Medical Research Council, Tygerberg, and [b]Department of Applied Medical Bioscience, University of the Western Cape, Bellville, South Africa

Introduction

The use of macaques has become popular in many areas of research, including human reproduction, although the reasons are mainly historical, not biological. However, there are several factors, such as the high costs of buying and shipping these monkeys from their country of origin, which limits their use in many developing countries. The vervet or African Green monkey, which is one of the two most important African primate species used in the laboratory, has emerged as an alternative local animal model in research on human reproduction [1]. This species has been used for reproductive studies at the Primate Unit of the South African Medical Research Council, and its basic sperm characteristics have been extensively defined at this facility. We present these sperm characteristics, relate the data to other non-human primate species and humans, and provide examples of the use of vervet monkeys as models for human male reproduction.

Materials and Methods

Sperm Characteristics

Semen Collection. Semen samples were collected from anaesthetized male vervet monkeys by rectal probe electroejaculation [8].

Concentration and Motility. The sperm concentration was determined using standard heamocytometer techniques. We determined motility by subjective and computerized methods. In the first, a drop of semen was placed between a microscope slide and coverslip, and motility was evaluated at 40× magnification. In the second method, 5 µl of semen was diluted in 1 ml of pre-warmed (37 °C) Hams-F-10. Details about viewing, recording and quantification of sperm motility using the Sperm Motility Quantifier have been described previously [5]. Motility parameters measured included curvilinear velocity (VCL), straightline velocity (VSL) and linearity (LIN).

Morphology. A thin semen smear was stained with Spermac stain according to the manufacturer's instructions.

pH. Merck universal indicator paper and Panpeha paper were used to determine the pH. The values obtained were compared to those determined with a pH meter on the same sample.

Table 1. Abnormal sperm morphology of vervet monkey

Sperm region	% abnormal
Head	1.64 ± 1.39
Midpiece	8.00 ± 7.29
Principal and end piece	29.32 ± 27.42

Table 2. pH determinations of seminal fluid from vervet monkeys

Type of analysis	pH value
Merck paper	8.75 ± 0.45
Panpeha paper	7.75 ± 0.50
pH meter	7.67 ± 0.20

Vitality and Acrosomal Integrity. The method employed for the preparation and monitoring of sperm vitality and acrosomal integrity with the use of Hoechst 33258 and fluorescein isothiocyante-peanut agglutinin (FITC-PNA) labeling techniques was essentially similar to that described before [7].

Examples of uses in reproductive research: Effects of estrogen and triterpenoid plant extract.

Results

Sperm Characteristics

Concentration and Motility. Sperm concentration was found to be 280.5 × 10^6/ml, progressive motility = 52%, and motion parameters were VCL = 110.69 µm/s, VSL = 75.52 µm/s, LIN = 58.92%.

Morphology. The normal morphology of vervet monkey spermatozoa followed the general human pattern, i.e. ovoid head with acrosomal and postacrosomal regions, and the midpiece which is slightly thicker than the tail. The quantitative evaluations of various morphological defects combined for each sperm region are presented in table 1.

pH. Comparison of mean pH values, obtained through three types of analysis methods, for ejaculates from 20 individuals is presented in table 2.

Vitality and Acrosomal Integrity. Results of these variables are presented in table 3. There were four distinct staining patterns of vervet monkey spermatozoa following labeling with FITC-PNA.

Uses in reproductive research: Affected certain sperm parameters.

Discussion

Concentration. Although there is a great variation in sperm concentration between species, the results indicate that sperm concentration of vervet monkeys is closer to that of humans than of other primate species, i.e. 419.43 × 10^6/ml for macaques (Harrison 1980) and 2,561 × 10^6/ml for chimpanzees [2].

Motility. Motion parameters of vervet monkey spermatozoa are closer to those of humans (VCL = 84 µm/s, VSL = 77.8 µm/s, LIN =

Table 3. Status of vervet monkey spermatozoa after staining with Hoechst 33258 and PNA-FITC

Status	%
Intact acrosome	53.58 ± 8.28
Patchy acrosome	8.00 ± 2.41
Equatorial staining	27.17 ± 5.32
No acrosome	11.33 ± 3.65
Live	78.75 ± 5.64

87 %) [6], than those of chimpanzees (VCL = 50 µm/s, VSL = 19 µm/s, LIN = 39 %) [2].

Morphology. Vervet monkey spermatozoa, like in other non-human primate species, have fewer head defects compared to humans [9].

pH. Our results suggests that where pH meter is not readily available, the Panpeha paper is most suitable for pH determinations of seminal fluid from vervet monkeys. Values obtained with these two methods are also within normal human range of 7.2–8.0 [10]. The Merck ColorpHast test paper is most suitable for seminal pH determinations in humans [3].

Acrosomal Integrity. The acrosomal staining procedure employed in this study demonstrates that a finite number of fluorescence patterns occur in vervet monkey spermatozoa, as in humans [7]. Therefore, it is speculated that the distribution of carbohydrate moieties on the surface of the spermatozoa, responsible for recognition of PNA lectin, is similar between human and vervet monkey spermatozoa. Our results show an abundance of spermatozoa with intact outer acrosomal membranes, which suggest that vervet monkey spermatozoa placed in the capacitating medium do not readily undergo spontaneous acrosomal loss in vitro. Similar phenomenon has also been observed with the squirrel monkey spermatozoa [4]. However, it has been observed that administrations of some aromatase inhibitors and the triterpenes to vervet monkeys lead to reductions in the above pattern.

Uses in reproductive research: The results show that both compounds are associated with reductions in certain sperm parameters of the vervet monkey spermatozoa.

Conclusion

Since values obtained for the variables investigated in this study fall within the normal range for adult fertile human males, this demonstrates the suitability of vervet monkeys in human reproductive studies.

References

1 Gombe S, Oduor-Okelo D, Else J: The potential of African mammals as new models for research in human reproduction; in Serio M, Martini L (eds): Animal Models in Human Reproduction New York, Raven Press, 1980, pp 345–358.
2 Gould KG, Young LG, Smithwick EB, Phythyon SR: Semen characteristics of the adult male chimpanzee *(Pan troglodytes)*. Am J Primat 1993;29: 221–232.
3 Haugen TB, Grotmol T: pH of human semen. Int J Androl 1998;21:105–108.
4 Kholkute SD, Lian Y, Roudebush WE, Dukelow WR: Capacitation and the acrosome reaction in squirrel monkey *(Saimiri sciureus)* spermatozoa evaluated by chlorotetracycline fluorescence assay. Am J Primatol 1990;20: 115–125.
5 Mdhluli MC, van der Horst G: The effect of oleanolic acid on sperm motion characteristics and fertility of male Wistar rats. Lab Anim 2002;36: 423–437.
6 Morales P, Overstreet JW, Katz DF: Changes in human sperm motion during capacitation in vitro. J Reprod Fertil 1988;83:119–128.
7 Mortimer D, Curtis EF, Miller RG: Specific labelling by peanut agglutinin of the outer acrosomal membrane of the human spermatozoa. J Reprod Fertil 1987;81:127–135.
8 Seier JV, Fincham JE, Menkveld R, Venter FS: Semen characteristics of vervet monkeys. Lab Anim 1989;23:43–47.
9 Sikka SC, Wang R, Kukuy E, Walker CF, Hellstrom WJ: The detrimental effects of electric current on normal human sperm. J Androl 1994;15:145–150.
10 World Health Organization (1992): WHO Laboratory Manual for the Examination of Human Semen and Sperm-Cervical Mucus Interaction, 3rd ed. Cambridge, Cambridge University Press, 1992.

8
In vitro Growth and Maturation of Oocytes in Human and Non-Human Primates

J.E.J. Smitz, R.G. Cortvrindt

Follicle Biology Laboratory, Academic Hospital-Free University Brussels, Brussels, Belgium

The Earliest Stages of Oocyte Growth in vivo and in vitro

Folliculogenesis is a process that takes many months in primates and in human [1]. Primordial follicles stored immediately under the ovarian capsula form the pool of immature gametes. These follicles are embedded into an interstitial tissue which is poorly vascularised and is considered to be metabolically inactive. From the moment that these primordial follicles are formed, which happens during fetal life in primates, some of these structures will start growth, initially characterised by a very slow transformation of the pregranulosa cell layer and an increase of the oocyte nucleus. The process of envel-oppement of oogonia in follicles is preceded by the initiation of the first meiotic division. Meiosis is than blocked in the prophase and a germinal vesicle is formed. There is a tremendous cell death in the fetal ovary: from the millions of oogonia present at mid gestation only a few hundred thousands primordial follicles remain. Oogonia not surrounded by granulosa cells and defect in the process of meiosis are correlated with the process of attrition. It is at present not completely understood by which physiological trigger(s) some of the resting follicles will be transformed into growing follicles. This process is called initial recruitment [2]. It has been proposed that either factors arising from the mesenchymal tissues (vascular or neuronal origin) or from the follicle itself could be initiator(s) of the growth process. Factors known to participate in this process are Kit Ligand (KL), basic fibroblast growth factor (bFGF), bone morphogenetic protein 7 (BMP-7) and Leukemia Inhibitory Factor (LIF) [3]. After recruitment into the growing pool the follicles will transform into a well defined series of successive follicular stages whereby a well defined compartmentalisation can be recognised [4]. The oocyte is separated from its intimate contact with pregranulosa cells by the zona pellucida but stays in contact with granulosa cells via specialised transzonal projections [5]. The basement membrane laid around the granulosa cells will be the barrier dividing the functional unit into an avascular (oocyte and granulosa) and vascularised part (theca externa). Acquisition of a theca layer is under the influence of stimuli originating in the oocyte (GDF-9). Formation of the theca interna cell layer is the hallmark of the early secondary follicle. This layer of cells from stromal origin will constitute a source of paracrine factors in the follicular unit. Subsequently a rich vascular network will be established around the follicle in a theca externa part. The whole process from resting primordial follicle up to a vascularised follicle takes months. It is recognised that the earliest stages of growth are passing through a very thightly programmed series of coordinated growth. The oocyte is undergoing impressive volume expansion during this period and experiments from Eppig's laboratory made clear that it is this cell that orchestrates the initial follicular growth stages [6].

Cryopreservation of Ovarian Tissue

For fertility preservation it is possible to cryopreserve ovarian cortical tissue [7]. Many researchers have proven that survival of the primordial follicle pool is approximately 70% after a freezing and thawing cycle [8, 9]. Theoretically this method could be used to preserve tissue from rare species and to enable young cancer patients to regain fertility after cure by a germ cell devastating chemo- or radiotherapy [10, 11]. It remains difficult to realise such a project: the ischemia that takes place at the moments of tissue prelevation and during tissue preparation, the revascularisation of the transplants and the choice of the best site for grafting remain actual questions. The majority of functional units are lost during the manipulations of the cortical tissue and not by cryodamage.

Transplantation of Ovarian Tissue

Experiments with retransplantation of tissues in humans have demonstrated that with the actual techniques the graft can function only during a couple of months [12, 13]. There might be excellent opportunities for refining the tranplantation procedures using primate models.

In vitro Culture of Ovarian Tissue

Research efforts have been pioneering techniques of in vitro culture of primordial follicles [14, 15]. Most of this work on large mammalians has been done on bovine, feline and porcine [16–18] and also on human [19]. There has been some work investigating use of fetal tissue from baboons [20]. The experimental results were quite concordant: in ovarian tissue from adult animals it is very difficult to obtain primordial follicles as functional units to culture [21]. Damage to the isolated cells occurs very easily and the units die off very quickly in culture [22]. A better approach seems to be tissue culture: different systems have been reported to allow follicle survival for a couple of weeks [19]. There are however not many data reported that illustrate a normally controlled growth comparable to an in vivo situation. Most often there is a distorded and uncoordinated growth between oocyte and granulosa cells [23]. There has been no demonstration that the oocytes from primordial follicles activated by in vitro conditions could resume meiosis. The experiments in human have shown that the content in follicles of the tissues obtained from clinical procedures in women in their thirtees is already strongly reduced [24–26]. Some first experiments using fetal ovarian tissue that contains enormous amounts of homogenously distributed primordial follicles could constitute a good source for this research. Primate models could function as an excellent source of tissue for this kind of research [20].

The Latest Stages of Oocyte Growth in vivo and in vitro

A couple of weeks or months (depending on the species) after the initial recruitment phase some of the follicles will have reached a stage at which they have become gonadotropin dependent for further growth and maturation [2]. During the period that precedes this gonadotropin dependent stage oocyte growth has occured concomitantly with follicular growth. During the oocyte's growth stage she acquires meiotic 'competence' allowing resumption of meiosis up to the second metaphase (MII) stage. This meiosis 'competence' (to reach MII) can occur in nearly all full-size oocytes. As long as the oocyte has not reached its full-size, it can undergo germinal vesicle breakdown (GVBD), but not reach MII. Oocyte growth is characterised by accumulation of RNA in the cytoplasm and the nucleolus. The RNA is stored in pools (specialised particles) and will be mobilised for protein synthesis during the period when there is transcriptional silencing of the genome i.e. from meiotic maturation until after the first embryonic cleavage divisions. Also protein stores, accumulated during the oocyte growth phase, are in the first place important for resumption of meiosis. The M-phase promoting factor (MPF), a protein composed of 2 subunits (cyclin B and $p34^{cdc2}$) has been well-studied [27] and its assembly to become active is regulated in a species-specific manner [28]. In humans the class of follicles that is recruited by the intercycle FSH rise belong to class V of the Gougeon classification [29]. These follicles have reached a diameter of 2 to 5 mm. In a natural cycle generally only a single follicle will grow up to a full maturation stage and ovulate. Interfering with the reduction of FSH during the follicular phase will cause many more follicles to grow up to a preovulatory stage. This principle is used quite efficiently in clinics to superovulate patients. The same techniques are used in several animal species for rodents up to non-human primates to increase the number of fertilisable oocytes for breeding purposes.

Studies in macacques have shown that the regulatory effects of gonadotropin on oocyte maturation and its capacity to sustain embryonic development is mediated by intraovarian factors [30]. Different intrafollicular environments will lead to varying responses of oocytes to similar gonadotropin exposures. Some studies in rodents and cattle suggest also an age-dependent effect of the oocyte donor on the developmental capacity of the oocyte [31, 32]. In rhesus monkey recent studies have shown that effects of FSH on oocyte developmental competence are modulated by animal age and breeding season [33].

Instead of using the principle of superovulation to obtain oocytes for assisted reproduction there is reviving interest in techniques that would allow collection of oocyte-cumulus-complexes from smaller follicle stages present at all moments of the cycle [34]. This principle could bring a reduction in the cost of treatment and in the complication rates observed after ovarian stimulation. In addition there is a pressing demand from the authorities to reduce the sharply increasing number of multiple pregnancies appearing in registries in the last few decades in Western countries [35, 15]. Although the IVM techniques have been used for many years in the bovine embryo production industry its application into the human model has not been very successful until today. After retrieval of the oocyte from the follicle, this cell is deprived of several crucial in vivo factors that play a role in its developmental competence [36]. These factors might originate in the systemic circulation or in the follicle wall inducing effects in interconnected granulosa-cumulus cells. It seems that the last days of the intrafollicular life time of the oocyte are crucial to acquire these factors. The mechanisms for sperm head penetration, decondensation of chromatin and the block to polyspermy are all acquired during the final days before ovulation [37].

As a general figure, the implantation potential of human embryo obtained after IVM is only half of that of an embryo obtained by conventional IVF procedures. The reduced rates of success with the IVM technique are due to multiple factors: the inability to precisely define the small follicles which are of good quality, the ignorance of whether there is a need for an appropriate pretreatment, the absence of an adequate culture procedure for cumulus-surrounded immature GV oocytes, the inability to monitor the cytoplasmic maturation stage of the cultured oocyte. Several actions of explorative work on improving oocyte competence can be undertaken such as: 1) defining a culture incubator that holds cumulus-granulosa-oocytes interconnected in vitro; 2) maintainance of meiotic arrest in vitro to permit cytoplasmic maturation to occur; 3) addition of endogenous factors known to improve cytoplasmic maturation to the culture medium; 4) addition of pharmacological compounds that promote cytoplasmic maturation; 5) co-culture designs etc.

Very few preliminar systematic research has been conducted in this area due to the scarce availability of human oocytes for ethical reasons. It is expected that work done to solve fundamental questions in several animal species such as the bovine and the rhesus model will help to push the development further in the interest of human ART. A non-human primate model is essential for understanding the regulation of primate oocyte maturation and to develop the techniques for primates [38]. Application of the IVM technology to non-human primates would also potentially increase numbers of viable embryos for biomedical research, including cloning and production of transgenic models for human disease [39].

References

1 Gougeon A: Regulation of ovarian follicular development in primates: facts and hypotheses. Endocr Rev 1996;17:121–155.
2 McGee EA and Hsueh AJW: Initial and cyclic recruitment of ovarian follicles. Endocrine Rev 2000;21:200–214.
3 Nilsson EE, Skinner MK: Growth and differentiation factor-9 stimulates progression of early primary but not primordial rat ovarian follicle development. Biol Reprod 2002;67:1018–1024.
4 Richards JS: Perspective: the ovarian follicle – a perspective in 2001. Endocrinology 2001;142:2184–2193.
5 Albertini DF, Combelles CM, Benecchi E, Carabatsos MJ: Cellular basis for paracrine regulation of ovarian follicle development. Reproduction 2001;121:647–653.
6 Eppig JJ, Wigglesworth K: Development of mouse and rat oocytes in chimeric reaggregated ovaries after interspecific exchange of somatic and germ cell components. Biol Reprod 2000;63:1014–1023.
7 Newton H, Aubard Y, Rutherford A, Sharma V and Gosden R: Low temperature storage and grafting of human ovarian tissue. Hum Reprod 1996; 11:1487–1491.
8 Oktay K, Nugent D, Newton H, Salha O, Chatterjee P, Gosden RG: Isolation and characterisation of primordial follicles from fresh and cryopreserved human ovarian tissue. Fertil Steril 1997;67:481–486.
9 Gook DA, Edgar DH, Stern C: Effect of cooling rate and dehydration regimen on the histological appearance of human ovarian cortex following cryopreservation in 1,2-propanediol. Hum Reprod 1999;14:2061–2068.
10 Gosden RG, Baird DT, Wade JC and Webb R: Restoration of fertility to oophorectomized sheep by ovarian autografts stored at −196 °C. Hum Reprod 1994;99:597–603.
11 Gosden RG, Mullan J, Picton HM, Yin H, Tan SL: Current perspective on primordial follicle cryopreservation and culture and culture for reproductive medicine. Hum Reprod Update 2002;8:105–110.
12 Oktay K, Karlikaya G: Ovarian function after transplantation of frozen, banked autologous ovarian tissue. N Engl J Med 2000;342:1919.
13 Radford JA, Lieberman BA, Brison DR, Smith AR, Critchlow JD, Russell SA, Watson AJ, Clayton JA, Harris M, Gosden RG, Shalet SM: Orthotopic reimplantation of cryopreserved ovarian cortical strips after high-dose chemotherapy for Hodgkin's lymphoma. Lancet 2001;357:1172–1175.
14 Smitz J, Cortvrindt R: The earliest stages of folliculogenesis in vitro. Reproduction 2002;123:185–202.
15 Picton HM: Oocyte maturation in vitro. Curr Opin Obstet Gynecol 2002; 14:295–302.
16 Wandji SA, Srsen V, Voss AK, Eppig JJ, Fortune JE: Initiation in vitro of growth of bovine primordial follicles. Biol Reprod 1996;55:942–948.
17 Jewgenow K: Role of media, protein and energy supplements on maintenance of morphology and DNA-synthesis of small preantral domestic cat follicles during short-term culture. Theriogenology 1998;49:1567–1577.
18 Hirao Y, Nagai T, Kubo M, Miyano T, Miyake M, Kato S: In vitro growth and maturation of pig oocytes. J Reprod Fertil 1994;100:333–339.
19 Hovatta O, Silye R, Abir R, Krausz T, Winston RM: Extracellular matrix improves survival of both stored and fresh human primordial and primary ovarian follicles in long-term culture. Hum Reprod 1997;12:1032–1036.
20 Wandji SA, Srsen V, Nathanielsz PW, Eppig JJ, Fortune JE: Initiation of growth of baboon primordial follicles in vitro. Hum Reprod 1997;12: 1993–2001.
21 Figueiredo JR, Hulshof SCJ, Van den Hurk R, Ectors FJ, Fontes RS, Nusgens B, Bevers MM, Beckers JF: Development of a new mechanical and enzymatical method for the isolation of intact preantral follicles from fetal, calf and adult bovine ovaries. Theriogenology 1993;40:789–799.
22 Van den Hurk R, Spek ER, Hage WJ, Fair T, Ralph JH, Schotanus K: Ultrastructure and viability of isolated bovine preantral follicles. Hum Reprod Update 1998;4:833–841.
23 Fortune JE, Cushman RA, Wahl CM, Kito S: The primordial to primary follicle transition. Mol Cell Endocrin 2000;163:53–60.
24 Faddy MJ, Gosden RG: A model conforming the decline in follicle numbers to the age of menopause in women. Hum Reprod 1996;11:1484–1486.
25 Lass A, Silye R, Abrams DC, Krausz T, Hovatta O, Margara R, Winston RM: Follicular density in ovarian biopsy of infertile women: a novel method to assess ovarian reserve. Hum Reprod 1997;12:1028–1031.
26 Cortvrindt R and Smitz J: Fluorescent probes allow rapid and precise recording of follicle density and staging in human ovarian cortical biopsies. Fertil Steril 2001;75:588–593.
27 Robert C, Hue I, McGraw S, Gagné D, Sirard MA: Quantification of cyclin B1 and $p34^{cdc2}$ in bovine cumulus-oocyte complexes and expression mapping of genes involved in the cell cycle by complementary DNA macroarrays. Biol Reprod 2002;67:1456–1464.
28 Yamashita M, Mita K, Yoshida N, Kondo T: Molecular mechanisms of the initiation of oocyte maturation: general and species-specific aspects. Prog Cell Cycle Res 2000;4:115–129.
29 Gougeon A: Dynamics of follicular growth in the human: a model from preliminary results. Hum Reprod 1986;1:81–87.
30 Schramm RD, Bavister BD: A macaque model for studying mechanisms controlling oocyte development and maturation in primates. Hum Reprod 1999;14:2544–2555.
31 Eppig JJ, Schroeder AC, O'Brien MJ: Developmental capacity of mouse oocytes matured in vitro: effects of gonadotrophic stimulation, follicular origin and oocyte size. J Reprod Fertil 1992;95:119–127.
32 Driancourt MA, Reynaud K, Smitz J: Differences in follicular function of 3-month-old calves and mature cows. Reproduction 2001;121:463–474.
33 Zheng P, Si W, Wang H, Zou R, Bavister BD, Ji W: Effect of age and breeding season on the developmental capacity of oocytes from unstimulated and follicle-stimulating hormone-stimulated rhesus monkeys. Biol Reprod 2001;64:1417–1421.
34 Chikazawa K, Araki S, Tamada T: Morphological and endocrinological studies on follicular development during the human menstrual cycle. J Clin Endocrinol Metab 1986;62:305–313.
35 Smitz J, Nogueira D, Cortvrindt R, de Matos DG: Oocyte in vitro maturation: state of the ART and basic requirements; in Gardner, Weissman, Howles, Shoham (eds.): Textbook of Assisted Reproductive Technology, London, Martin Dunitz LTD, 2001, Chapter 9, pp 107–137.
36 Bevers MM, Dieleman SJ, van der Hurk R, Izadyar F: Regulation and modulation of oocyte maturation in the bovine. Theriogenology 1997;47: 13–22.

37 Trounson A, Anderiesz C, Jones G: Maturation of human oocytes in vitro and their developmental competence. Reproduction 2001;121:51–75.
38 Schramm RD, Paprocki AM, Bavister BD: Features associated with reproductive ageing in female rhesus monkeys. Hum Reprod 2002;17:1597–1603.
39 Schramm RD, Paprocki AM: Birth of rhesus monkey infant after transfer of embryos derived from in vitro matured oocytes: short communication. Hum Reprod 2000;15:2411–2414.

9

Non-Human Primates as a Model for Reproductive Aging and Human Infertility

C.A. Brenner[a–c], *S.M. Nichols*[a, b], *E.S. Jacoby*[a, b], *B.D. Bavister*[a, c]

[a] Department of Biological Sciences, University of New Orleans, La., [b] Audubon Center for Research of Endangered Species, New Orleans, La.; [c] The Jones Institute for Reproductive Medicine, Department of Obstetrics and Gynecology, Eastern Virginia Medical School, Norfolk, Va., USA

Deliberate harvesting of human oocytes and creation of human embryos for scientific research is surrounded by ethical, political and practical considerations. Typically, to obtain human material for experimental studies, research proposals must initially undergo intense scrutiny by an internal review board (IRB). In addition, most studies must be performed using discarded in vitro fertilization (IVF) material that has prematurely arrested development or appears to be of substandard quality. Experimental outcomes obtained with such material may not reflect normal processes of fertilization and pre-implantation development.

Human reproductive research is also associated with high costs not experienced by researchers using other animal models. As a result, many studies attempting to define the molecular mechanisms of reproduction utilize models that are economically feasible. However, data obtained from certain alternative models, such as the mouse, may not be suitable for extrapolation to human reproduction. There are fundamental differences between mice and primates. For example, centrosomal inheritance during fertilization is paternal in primates, whereas in rodents, the centriole is maternally-derived [1]. In primates, chorionic gonadotropin (CG) secretion by embryos is needed to maintain pregnancy, but not in rodents.

The ideal model for human reproductive studies, outside of humans themselves, would possess similar physiological, genetic and cytoplasmic components. Non-human primates are our closest phylogenetic relatives, and their embryo morphology and pre-implantation development closely resembles that of humans. There are convincing reasons why non-human primates such as the baboon or rhesus monkey should be used for testing novel procedures, including in vitro oocyte maturation (IVM), oocyte and blastocyst cryopreservation, development of new culture media and novel approaches to infertility treatment. Through assisted reproductive technologies (ART) we can obtain presumptively normal non-human primate embryos for scientific study and for testing safety and efficacy of new techniques and procedures prior to their use in IVF clinics.

The practice of assisted reproduction has changed dramatically over the last decade. Physicians are more adept at ovarian stimulation protocols, managing hyper-stimulation and selecting the optimal day for administering hCG. In vitro fertilization laboratories are better equipped, staffed and more knowledgeable about human oocyte and embryo development. It is evident that a dramatic decline in fertility begins in female patients past their middle thirties [2, 3]. This 'reproductive senescence', which is marked by a sharp decline in embryo implantation and pregnancy rates, may be attributed to reduced number of oocytes obtained at retrieval or number of embryos produced, and also, to initial quality of oocytes.

In this article, we will review the status of reported genetic abnormalities in human oocytes and embryos and conjecture how non-human primates including the baboon are necessary models for reproductive senescence and human infertility. It is critically important to understand the mechanisms of cytoplasmic and nuclear anomalies underlying reproductive senescence so that assisted reproductive technology can be employed in older patients. However, any new technology must be shown to be safe and effective before it is applied to the infertile patient.

Chromosomal Status

Chromosomal aneuploidies are a major cause of fetal abnormalities, with an incidence of 21% in spontaneously aborted pregnancies. Of these, trisomies of chromosomes 21, 18, 15 and 13 may account for over 50% of the karyotypically abnormal miscarriages. In contrast to single gene defects, aneuploidies are the only risk factor definitively known to be associated with maternal age, increasing from 2% in recognized pregnancies of women 25 years old to over 19% in women 40 years or older [2]. Use of fluorescent in situ hybridization (FISH) to screen for aneuploidies of chromosomes X, Y, 18, 13, and 21 in individual blastomeres of human embryos prior to transfer may potentially reduce the risk of miscarriage or reduce the number of offspring possessing trisomies. These screenings become increasingly important in older IVF patients [3].

Pre-implantation genetic diagnosis (PGD) of aneuploidy may increase the pregnancy rate in IVF patients of advanced maternal age by allowing for the selection of only normal embryos for transfer. The causes of the decline in implantation rates observed with increasing maternal age are still under debate. Nevertheless, the high implantation rate obtained with oocyte donation from younger women strongly suggests that maternal age of oocytes is a key factor in reproductive senescence [4]. Pre-implantation genetic diagnosis of aneuploidy becomes increasingly advantageous when we consider that in the United States, 52% of IVF stimulation cycles are performed in women 35 years of age or older (ASRM-SART, 1998).

Fluorescent in situ hybridization allows chromosome enumeration to be performed on the interphase cell nuclei without the need for preparing complicated metaphase spreads. This technology has been applied to PGD of common aneuploidies using either human blastomeres from 2- to 16-cell stage embryos or oocyte polar bodies [5]. Currently, probes for chromosomes X, Y, 13, 14, 15, 16, 18, 21, and 22 are being used simultaneously, demonstrating the potential to detect 70% of the aneuploidies that result in spontaneous abortion [3, 5].

A major problem associated with these approaches is that diagnosis of aneuploidy during pre-implantation stages of embryonic development does not take into account errors that may occur in subsequent divisions, resulting in mosaicism. Furthermore, implantation rate may be reduced by the embryo biopsy procedure. Although many studies on embryo biopsy do not show a negative effect on embryo development to blastocyst [5, 6], pregnancy outcomes after biopsies have yet to be assessed. Analysis of whole chromosome sets using techniques such as comparative genomic hybridization (CGH) of single cells can potentially provide more clinically useful information [6].

In order for non-human primates to become indispensable models for analyzing the etiology of chromosome anomalies, as well as for developing new technologies to overcome these deficiencies, we must place much greater emphasis on the use of these animals for reproduction research. Further, we must show that technologies already in place in humans (such as FISH) can be readily achieved in non-human primate oocytes and embryos, so that we can extrapolate information obtained from them to the human condition.

In addition to aneuploidy analysis, recent studies have led to improvements in understanding molecular mechanisms controlling chromosomal segregation. The current model of eukaryotic cell cycle regulation suggests that there is an oscillating biochemical clock regulated by surveillance systems called cell cycle checkpoints. The spindle assembly checkpoint modulates the timing of anaphase initiation in response to improper alignment of chromosomes at the metaphase spindle. If defects are detected, a signal is transduced to halt further progression of the cell cycle until correct bi-polar attachment to the spindle is achieved. MAD2 and BUB1 genes encode conserved kinetochore-associated proteins believed to be components of the checkpoint regulatory pathway. A failure in this surveillance system could lead to genomic instability that may underlie increased incidence of aneuploidy in gametes of older women. To explore this possibility, the copy number of MAD2 and BUB1 transcripts has been determined in human oocytes at various stages of maturation using a quantitative, real-time rapid cycle fluorescence RT-PCR method [7, 8]. The results suggest that these messages degrade in oocytes from older women. This degradation may impair checkpoint function in older oocytes and be a contributing factor to age-related aneuploidy [9]. The presence of MAD2 at kinetochores of normal, metaphase II mouse oocytes has recently been confirmed (Blaszczyk et. al., unpublished). Further protein analysis is in progress to determine whether the lack of cell cycle checkpoint proteins play a pivotal role in the etiology of age-related aneuploidy. The ability to analyze whole cohorts of oocytes from normal, fertile monkeys can be of enormous help in these analyses.

The notion that chromosomal abnormalities in human oocytes are the result of premature chromatid separation has led to some interesting hypotheses. In the case of premature separation of sister chromatids, a series of proteins called cohesins may be missing or defective (Brenner, unpublished results). Cohesins bound to chromosomal sequences flanking the centromeres prevent sister chromatids from completely unzipping and pull back sister centromeres that have prematurely split [10]. Cohesins, consequently, play a central role in the generation of adhering tension between microtubules and sister chromatids at the centromeres. Additionally, degradation of cohesin molecules by proteases is essential for chromosome stability. The family of proteins that cleaves the chromosome cohesins is called separins [10]. Many of these proteins have been characterized only in yeast, and it remains to be seen if human homologues are implicated in the processes associated with aneuploidy in human oocytes. Since it may be difficult to conduct such research studies in humans, it is essential to explore non-human primate models that exhibit similar chromosomal aneuploidies resulting from reproductive aging.

Mitochondria in Oogenesis and Aging

Mitochondria play a key role in early development, though one that is far from being fully understood [11–13], hence there is a need to study them in oocytes and pre-implantation embryos. There are striking rearrangements of mitochondrial localization during fertilization indicating a fundamental role for mitochondria in embryogenesis. While the functional significance of mitochondrial re-localizations is unknown, they are clearly associated with developmental competence in rodents [14, 15] and domestic species [16–18]. In rhesus monkeys, the mitochondrial localization profile appears to be correlated with embryo development after IVF. However, this remains to be confirmed. A positive relationship has been inferred between peri-pronuclear mitochondrial accumulation in human oocytes and their developmental capacity [19, 20], but the extent of the relationship is not known. The kinetics of this peri-pronuclear mitochondrial accumulation needs to be studied to determine if it shows an age-related decline.

Progressive damage to mitochondrial DNA (mtDNA) during life is thought to contribute to the aging process. Point mutations may accumulate with age in the human mtDNA control region for replication in human fibroblasts, muscle and cardiac cells. If a similar phenomenon occurs in aging oocytes within the ovary, this would raise significant questions concerning oocyte quality. It is significant that muscle- and fibroblast-specific aging-related mtDNA mutations occur at or near the site of DNA attachment to the inner mitochondrial membrane, which is a likely site of nucleo-mitochondrial interactions and is also involved in mitochondrial metabolic activity. Hence, the ability of mitochondria to provide energy for fertilization and subsequent development may decline with age. Therefore, the important biological question is whether an increasing percentage of mtDNA mutational load corresponds to poor oocyte quality, poor embryo development and implantation failure in the older patient. Perhaps a minimum number of functionally active mitochondria is required to sustain the resumption of meiosis, fertilization, pre-implantation and early fetal development. This needs to be examined in a non-human primate model so that oocytes and embryos can be examined in a detailed manner.

Murine versus Non-Human Primate Models to Evaluate Reproductive Senescence

Murine models have played an important role in our understanding of the differences between male and female meiosis. Although economical to produce, intrinsic qualities of murine gametes and embryos make it difficult to relate findings to human fertility. Murine oocytes and pre-implantation embryos exhibit a low naturally-occurring frequency of aneuploidy compared to their human counterparts. Unexpectedly, when meiotic errors are produced in available mouse mutants, male meiosis grinds to a halt while female meiosis continues to progress [21]. This finding is in contrast to primates, in which oogenesis possesses much more stringent quality controls than spermatogenesis. In addition, murine gametes simply do not behave in the same manner as primate gametes when exposed to various advanced reproductive technologies. Therefore, it is imperative that the natural occurrence of aneuploidy is further investigated in non-human primates so that we may elucidate the molecular and cellular machinery that causes these meiotic errors in oocytes during reproductive senescence in humans. Clues about the etiology of this anomaly are most likely to come from studies with non-human primate models like the baboon and rhesus monkey.

References

1 Hewitson L, Schatten G: The use of primates as models for assisted reproductive technologies. Reprod Biomed Online 2002;5:50–55.
2 Warburton D: Genetic factors influencing aneuploidy frequency. Basic Life Sci 1985;36:133–148.
3 Munne S, Cohen J: Chromosome abnormalities in human embryos. Hum Reprod Update 1998;4:842–855.
4 Navot D, Drews MR, Bergh PA, Guzman I, Karstaedt A, Scott RT, Garrisi GJ, Hoffmann GE: Age-related decline in female fertility is not due to diminished capacity of the uterus to sustain embryo implantation. Fertil Steril 1994;61:97–101.
5 Verlinsky Y, Cieslak L, Ivakhnenko V, Evsikov S, Wolf G, White M, Lifshez A, Kaplan B, Moise J, Valle J, Ginsberg N, Strom C, Kuliev A: Pre-pregnancy genetic testing for age-related aneuploidies by polar body analysis. Genet Test 1997–98;1:231–235.
6 Wells D, Delhanty JD: Comprehensive chromosomal analysis of human pre-implantation embryos using whole genome amplification and single cell comparative hybridization. Mol Hum Reprod 2000;6:1055–1062.
7 Steuerwald N, Cohen J, Herrera RJ, Brenner CA: Analysis of gene expression in single oocytes and embryos by real-time rapid cycle fluorescence monitored RT-PCR. Mol Hum Reprod 1999;5:1034–1039.
8 Steuerwald N, Cohen J, Herrera RJ, Brenner CA: Quantification of mRNA in single oocytes and embryos by real-time rapid cycle fluorescence monitored RT-PCR. Mol Hum Reprod 2000;6:448–453.
9 Steuerwald N, Cohen J, Herrera R, Sandalinas M, Brenner CA: Association between spindle assembly checkpoint expression and maternal age in human oocytes. Mol Hum Reprod 2001;7:101–109.
10 Nasmyth K, Jan-Michael P, Uhlmann F: Splitting the chromosome: Cutting the ties that bind sister chromatids. Science 2000;288:1379–1384.
11 Barnett DK, Bavister BD: What is the relationship between the metabolism of preimplantation embryos and their development in vitro? Molec Reprod Dev 1996;43:105–133.
12 Bavister BD, Squirrell JM: Mitochondrial distribution and function in eggs and early embryos. Hum Reprod 2000;15(suppl 2):189–198.
13 Barnett DK, Kimura J, Bavister BD: Translocation of active mitochondria during hamster preimplantation embryo development studied by confocal laser scanning microscopy. Dev Dynamics 1996;205:64–72.
14 Ludwig TE, Squirrell JM, Palmenberg AC, Bavister BD: Relationship between development, metabolism and mitochondrial organization in 2-cell hamster embryos in the presence of low levels of phosphate. Bio Reprod 2001;65:1648–1654.
15 Muggleton-Harris AL, Brown JJ: Cytoplasmic factors influence mitochondrial reorganization and resumption of cleavage during culture of early mouse embryos. Hum Reprod 1988;3:1020–1028.
16 Hyttel P, Niemann H: Ultrastructure of porcine embryos following development in vitro versus in vivo. Mol Reprod Dev 1990;27:136–144.
17 Krisher RL, Squirrell JM, Bavister BD: Mitochondrial distribution in bovine oocytes matured in vitro is affected by maturation medium and is related to developmental competence. Reproduction, submitted.
18 Krisher RL, Bavister BD: Responses of oocytes and embryos to the culture environment. Theriogenology 1998;49:103–114.
19 Van Blerkom J, Davis P, Alexander S: Differential mitochondrial distribution in human pronuclear embryos leads to disproportionate inheritance between blastomeres: Relationship to microtubular organization, ATP content and competence. Hum Reprod 2000;15:2621–2633.
20 Sathananthan AH, Trounson AO: Mitochondrial morphology during preimplantational human embryogenesis. Hum Reprod 2000;15(suppl 2):148–159.
21 Hunt PA, Hassold TJ: Sex matters in meiosis. Science 2002;296:2181–2183.

10

Ovarian Stimulation, Egg Aspiration, in vitro Fertilization and Embryo Transfer in the Baboon *(Papio anubis)*: A Pilot Project at the Institute of Primate Research, Nairobi, Kenya

Thomas M. D'Hooghe[a,b], Carl S. Spiessens[b], Daniel C. Chai[a], Peter G. Mwethera[a], Aloys O. Makokha[a], Jason M. Mwenda[a]

[a] Institute of Primate Research, Karen, Nairobi, Kenya;
[b] Leuven University Fertility Center, Department of Obstetrics and Gynecology, University Hospital Gasthuisberg, Leuven, Belgium

Introduction

In women, ovarian stimulation is performed using urinary or recombinant gonadotrophins including HMG (human menopausal gonadotrophin), a mixture of FSH and LH derived from urine of postmenopausal women, FSH (either recombinant FSH, or FSH derived from urine). To prevent premature ovulation, this treatment is combined with the administration of an LHRH agonist (buserelin, goserelin) or LHRH antagonist (cetrotide, ganirelix, antide). As a result, in women, 8–10 mature eggs can be collected per oocyte aspiration [1].

In women and in nonhuman primates, the application of ART requires a large and consistent supply of oocytes. Oocytes have been recovered in vivo from unstimulated primates such as baboons, marmosets, chimpanzees, rhesus and pig-tailed macaques during spontaneous menstrual cycle [2–4]. However, the collection of oocytes from nonstimulated animals is tedious and of low yield. Several methods for stimulating the growth and maturation of multiple preovulatory follicles and their oocytes in nonhuman primates have been reviewed [5–8].

In the development of IVF in non-human primates, human gonadotropins have been used, but their effectiveness has been hampered by immunopathological reactions with the production of neutralizing antibodies against human gonadotropins of urinary origin (reviewed by Ouhibi et al., 2000 [9]). However, it has been reported [5] that the presence of these antibodies did not interfere with conception or maintenance of pregnancy in rhesus monkeys. Recently, it has been reported that circulating antigonadotropin antibodies were absent following one cycle of ovarian with recombinant hFSH/LH and recombinant HCG in rhesus monkeys [10]. After 3 cycles of ovarian stimulation with these products, the majority of rhesus monkeys had anti-hFSH (60%) and anti-hCG (70%) antibodies, but still had mature fertilizable oocytes [10]. It was concluded that in rhesus monkeys up to 3 follicular stimulation cycles per macaque can be achieved using recombinant human gonadotropins without major impact on oocyte maturation at collection or fertilization in vitro [10].

In baboons, it has been reported that ovarian stimulation with hMG and hCG results in multifollicular growth in 11 of 19 animals studied, but no follicles were observed during the second stimulation cycle, probably due to neutralizing antibodies against human FSH, LH and HCG , suggesting that baboons can only be stimulated with human hMG during one cycle [11]. Recent experience at the Institute of Primate Research demonstrated that baboons can be superovulated by injection of 1 ampule human menopausal gonadotropin, 75 IU (Metrodin, Serono, Rome, Italy) during 5 days from cycle day 7, followed by 5,000 IU chorionic gonadotropin (Profasi, Serono, Rome, Italy) 36 h after the last gonadotropin injection. On average, 30 ± 5 eggs were harvested per baboon [6].

The goal of this pilot study was to test the hypothesis that eggs harvested from female baboons after ovarian stimulation with human gonadotrophins can be fertilized in vitro with baboon sperm and develop into embryos that can be transferred in the same baboon using a transcervical approach, as in women.

Materials and Methods

Baboons

All baboons (PAN 2802, PAN 2838, PAN 2574, PAN 2652) were be of proven fertility in the wild (pendulous nipples at the time of capture), did not have any previous surgeries or diseases, were in good health and had regular menstrual cycles. Cycle day 1 was defined as the first day of menstrual bleeding.

Ovarian Stimulation

The four baboons were divided into two groups (2 animals per group) and given a different regime of ovarian stimulation, modified according to a previously published protocol [12]. Briefly, after pituitary downregulation with triptorelin (Decapeptyl, Ipsen), baboons were daily treated with human recombinant FSH IM (Gonal-F, Serono, Geneva, Switzerland) or purified human urinary FSH and LH IM (Menopur, Ferring) for 14 (group 1) or 11 (group 2) days, followed by intramuscular injection of HCG 2500 IU (Profasi, Serono, Geneva Switzerland) and laparoscopic egg retrieval 26 h later.

Anesthesia, Ovarian Ultrasound and Laparoscopic Egg Retrieval

Baboons were anesthetized, transferred from the cages to the surgery, weighed, shaved and prepared for laparoscopy as described previously [6, 13]. A transabdominal ultrasound was performed prior to laparoscopy to document ovarian size and the number/size of ovarian follicles. This was done with a vaginal ultrasound probe (Aloka 500, UST-981P-5, 5 MHz, Corometrics Medical Systems, Inc.), since the abdominal probe (UST-981P-5) was too large for the lower abdomen of these adult female baboons. Print-outs were made (Sony, Video Graphic Printer, UP-811)

The laparascopic egg retrieval was modified according to a previous protocol in baboons [6]. Briefly, an ovum aspiration needle also used in women (COOK® K-OPSD-1725-ET, Cook, Australia) was connected via a silicone rubber tubing to a trap. This trap, a 15-m Corning centrifuge tube, was connected via a 2-foot tube to a footswitch aspiration pump (Mefar Surgical Suction Pump, M20, Mefar, Bovezzo, Italy). A negative suction pressure of –50 kPA in the first group, as reported previously [6]. Based on the egg recovery results in the first group (many eggs without cumulus, see Results), the negative suction pressure was reduced to –25 kPA and to –15 kPA in the second group.

After egg retrieval, the pelvic cavity was rinsed using a Normal Saline solution.

Sperm Collection

Sperm was obtained from a male baboon of proven fertility through electro-ejaculation. At ejaculation the sperm was collect in 2 ml of prewarmed (37 °C) sperm buffer (Cook, Australia). The sperm suspension was kept at 37 °C for 30 min to allow liquification. After this incubation the coagulum was removed and the sperm suspension was centrifuged for 10 min at 400 *g* and the pellet was resuspended in 0.5 ml of fertilization medium (Cook, Australia).

In the first group, this sperm suspension was used for either a swim up procedure or for a density gradient centrifugation.

For the swim up procedure the sperm was put below 1.5 ml fertilization medium and incubated for 45 min at 37 °C. After the incubation the upper layer was collected, its sperm motility was evaluated and used for insemination of the oocytes.

For the gradient centrifugation the sperm suspension was placed on a discontinuous two layer Percoll gradient (45%/90%) and centrifuged for 20 min at 400 *g*. The pellet was collected and washed twice in fertilisation medium. Finally the sperm were resuspended in fertilization medium and, after an incubation of 30 min at 37 °C, the sperm quality was evaluated.

This final sperm suspension was used for insemination.

In the second group, the same procedure was followed, but sperm were also activated using pentoxyfylline (3.6 m*M*) incubation for 15 min just before insemination. The sperm suspension was washed by centrifugation to remove the pentoxyfylline. The sperm were resuspended in fresh fertilization medium and immediately used for insemination.

Collection of Oocytes

The retrieved oocytes were washed in oocyte wash buffer (Cook, Australia) and incubated in fertilization medium.

In the first group, the oocytes were inseminated after 4–6 h of incubation. The insemination was performed using 400,000 motile sperm per oocyte. The presence of 2 pronuclei was checked after overnight incubation.

In the second group, the oocytes were first maturated in vitro matured during 18 to 20 h. At that time, oocyte maturity was evaluated, insemination was performed and the inseminated oocytes were incubated for 24 h. The presence of 2 pronuclei was checked 24 h after insemination.

Embryo Culture and Transfer

Fertilized oocytes and embryos were cultured in cleavage medium (Cook, Australia).

Intrauterine embryo transfer was performed as in humans, using a commonly used soft-tip embryo transfer together without the need for additional cannula and/or stylet (Soft Trans Universal K-Soft-5001, Cook, Australia). The uterus was easily accessed up to 5 cm beyond the external ostium of the cervix. Since resistance was met at that point, the catheter was retracted 1 cm. The localization of the embryo transfer catheter in the uterus was confirmed by ultrasound. It was not necessary to do a laparoscopic stretching of the uterus [13] to facilitate the embryo transfer procedure.

Diagnosis of Pregnancy

The diagnosis of pregnancy was made using clinical criteria (amenorrhea with flat and red colored perineum) with the intention to perform an obstetric ultrasound 6 weeks after embryo transfer.

Results

Baboons

All baboons supported the treatment well, but developed a strongly inflated perineum at the time of egg retrieval. The degree of inflation was higher than usually observed in baboons at the end of the follicular phase. One baboon (Pan 2652) had important infected lacerations in the perineal area at the time of egg retrieval, probably due to injury. The laparoscopic egg retrieval was performed without complications in all baboons.

Oocytes and Sperm

In the first group, 36 and 29 eggs were collected from Pan 2838 and Pan 2802, respectively.

Nearly all eggs were 'naked' at microscopic inspection, i.e. they lacked the characteristic cumulus granuloma cells. The sperm was prepared using density gradient centrifugation. The sperm suspension used for insemination showed 73% progressive motile sperm. The evaluation of oocyte maturity at the time of fertilization control (18–20 h after aspiration) showed respectively 12/36 and 5/29 mature oocytes in Pan 2838 and Pan 2802, respectively. One oocyte of each baboon showed 2 pronuclei. The sperm motility at the time of the fertilization control was below 10%. None of the 2 fertilized oocytes developed into an embryo.

In the second group, 36 and 32 eggs were collected from Pan 2652 and Pan 2574, respectively. Overnight in vitro maturation resulted in 16/36 and 18/32 mature oocytes, respectively. Insemination was performed using sperm from a swim up procedure activated with pentophylline. Two pronuclei were observed in 1 oocyte from Pan 2574. Further culture resulted in a 2 cell stage embryo suitable for embryo transfer.

Embryo Transfer

One baboon from the second group, PAN 2574, received an embryo in a 2 cell stage using transcervical transfer 3 days after egg retrieval. This baboon started her menstruation after a prolonged luteal phase of 20 days, but without the red perineal discoloration typical for a clinical pregnancy in baboons. The duration of the luteal phase in the other 3 baboons that did not receive an embryo transfer was considerably shorter.

Discussion

This pilot study in baboons has shown the feasibility of pituitary desensitization, ovarian stimulation using human gonadotrophins, laparoscopic egg aspiration, egg maturation in vitro, sperm preparation, in vitro fertilization, embryo cleavage and autologous transcervical uterine transfer.

The swim-up procedure for sperm preparation resulted in a higher % motile sperm at insemination and a higher sperm survival (motility) 24 h after insemination when compared to the sperm density gradient centrifugation. This higher motility should maximize the chance of fertilization. Whether sperm should be activated using pentoxiphylline remains to be investigated in a larger controlled study.

The relative low proportion of mature eggs (less than 50%), considerably lower than that observed in women (normally 80%), is a matter of concern, and may be related to the protocol for ovarian stimulation (dose too high?). Further studies are also needed to test the hypothesis that extension of the time interval between HCG injection and egg retrieval from 26–36 h, like in humans, may increase the proportion of mature fertilizable oocytes.

The aspiration pressure during egg retrieval also was too high in the first group, where nearly all eggs retrieved in the first group did not have a cumulus. In the second group, aspiration pressure was reduced and more eggs had a cumulus, but further reduction of aspiration pressure may further benefit the quality of eggs.

To the best of our knowledge, this is the first report showing that autologous transcervical embryo transfer is possible in baboons. In rhesus monkeys, embryo transfers only lead to a pregnancy when 4–8-cell embryos are transferred via the oviduct to heterologous hormonally prepared recipient rhesus monkeys [14]. Similarly, laparoscopic oviductal transfer was performed in a study in baboons [15].

The baboon receiving embryo transfer in our study did not become clinically pregnant, but a chemical pregnancy cannot be ruled out since the luteal phase lasted longer than in the other 3 baboons that had not received an embryo transfer. An ongoing pregnancy after laparoscopic oviductal embryo transfer has been published in only one report [15] so far, to the best of our knowledge. In this study, in vitro maturation of 22 preovulatory eggs during 24 h, followed by in vitro fertilization with 0.36 million progressive motile sperm per ml, resulted in 19 fertilized eggs [15]. Laparoscopic oviductal transfer of 3 pronuclear zygotes and 1 2-cell embryo led to a viable pregnancy and livebirth in 1 out of 2 baboons treated [15].

This pilot study provides the basis for additional studies the optimize all steps of the clinical and laboratory procedures related to assisted reproduction in the baboon. This is important, since the establishment of a reliable system of assisted reproduction in baboons may offer a unique preclinical model to study problems related to human reproduction (i.e. embryo implantation in vivo and in vitro) and to embryonic stem cell development and therapy.

References

1 Spiessens C, D'Hooghe TM, Wouters E, Meuleman C, Vanderschueren D: Alpha-glucosidase activity in seminal plasma: Predictive value for outcome in intrauterine insemination and in vitro fertilization. Fertil Steril 1998;69: 735–739.

2 Clayton O, Kuehl TJ: The first successful in vitro fertilization and embryo transfer in a nonhuman primate. Theriogenology.

3 Cranfield MR, Bavister BD, Boatman DE, Berger NG, Schaffer N, Kempske SE, et al: Assisted reproduction in the propagation management of the endangered lion-tailed macaque *(Macaca silenus)*; in Wolf DP, Stouffer RL, Brenner RM (eds): In vitro fertilization and embryo transfer in primates. New York, Springer, 1993, pp 331–348.

4 Pope CE, Dresser BL, Chin NW, Liu JH, Loskutoff NM, Behnke EJ, et al: Birth of a Western lowland gorilla *(Gorilla gorilla gorilla)* following in vitro fertilization and embryo transfer. Am J Primatol 1997;41:247–260.

5 Iliff SA, Molskness TA, Stouffer RL: Anti-human gonadotropin antibodies generated during in-vitro fertilization (IVF)-related cycles: Effect on fertility of rhesus macaques. J Med Primatol 1995;24:7–11.

6 Mwethera PG, Makokha A, Chai D: Fertilin β peptides inhibit sperm binding to zona-free eggs in a homologous baboon in vitro fertilization system. Contraception 1999;59:131–135.

7 Wolf DP, Thomson JA, Zelinski-Wooten MB, Stouffer RL: In vitro fertilization-embryo transfer in nonhuman primates: The technique and its application. Mol Reprod Dev 1990;27:261–280.

8 Stouffer RL, Zelinski-Wooten MB, Aladin Chandrasekher Y, Wolf DP: Stimulation of follicle and oocyte development in macaques for IVF procedures; in Wolf DP, Stouffer RL, Brenner RM (eds): In vitro fertilization and embryo transfer in primates. New York, Springer, 1993, pp 124–141.

9 Ouhibi N, Zelinski-Wooten MB, Thomson JA, Wolf DP: Assisted fertilization and nuclear transfer in nonhuman primates; in Contemporary Endocrinology. Towata, Humana Press, 2001.

10 Zelinski-Wooten MB, Hutchison JS, Trinchard-Lugan I, Hess DL, Wolf DP, Stouffer RL: Initiation of periovulatory events in gonadotrophin-stimulated macaques with varying doses of recombinant human chorionic gonadotrophin. Human Reprod 1997;12:1877–1885.

11 McCarthy TJ, Fortman JD, Boice ML, Fazleabas AT, Verhage HG: Induction of multiple follicular development and superovulation in the olive baboon, *Papio anubis*. J Med Primatol 1991; 20:308–314.

12 Caperton L, Brindle E, Fujita M, O'Connor KA, Atencio J, Jackson E, Rice K, Leland M, Carey KD, Eddy C, McCarrey JR: Serum and urinary hormonal analytes for the prediction of ovulation in baboons. Biol Reprod 2003;68(suppl 1):281.

13 D'Hooghe TM, Bambra CS, Cornillie FJ, Isahakia M, Koninckx PR: Prevalence and laparoscopic appearance of spontaneous endometriosis in the baboon *(Papio anubis, Papio cynocephalus)*. Biol Reprod 1991;45:411–416.

14 Hewitson, L, Dominko, T, Takahashi D, et al: Unique checkpoints during the first cell cycle of fertilization after intracytoplasmic sperm injection in rhesus monkeys. Nature Med 1999;5:431–433
15 Clayton O, Kuehl TJ: The first successful in vitro fertilization and embryo transfer in a nonhuman primate. Abstract Proc Xth Annu Conf Int Embryo Transfer Society, San Antonio, 1984. Theriogenology 1984;21:228.

11
Primate Cloning and Stem Cells – Options and Impacts

J.P. Hearn
Department of Physiology, University of Sydney, Sydney, Australia

Introduction

Two research advances published 5 years ago opened an opportunity for a revolution in many areas of biomedicine and health. The first was the demonstration of 'reproductive cloning' with the birth of Dolly. The second was the isolation and characterisation of human embryonic stem cells. The papers, published in 1996 by Wilmut et al. and in 1998 by Thomson et al., respectively, triggered a remarkable rethink of the limits to cell potency differentiation in mammals. The potential applications could touch many aspects of life including agriculture and food, biomedicine and health, environment and conservation. (Note: The points discussed below refer primarily to primates and to mammals, but the knowledge base derives from 50 years of research in amphibians and rodents.)

Dolly is now dead, possibly prematurely but apparently due to respiratory disease. Since her birth, the number of mammalian species cloned by cell nuclear transfer (CNT) has been expanded to rodent, bovine and feline species, but the failure rate through early or late termination of pregnancy and neonatal death is still very high and repetition of the procedure continues to be difficult and unpredictable. The reason suggested for embryonic or neonatal death is given as inadequate or damaged organs or systems. We have much to learn about the reprogramming of the adult nucleus by the cytoplasm of the egg, which is at the core of the cloning process. The speed at which the process is carried out evidently causes substantial damage – with 1 in every 20 genes claimed to be affected. Cloning is not a routine or an easily repeatable procedure [Rhind et al., 2003].

The whirlwind of world-wide debate that resulted from publication of the above two advances was in stark contrast to the minor ripple that radiated from the earlier and similar achievement of embryonic stem cell derivation in the rhesus in 1995 and the marmoset [Thomson et al., 1996], although very similar protocols had been developed for these two species before they were taken to the human. The work in these two primate systems was the result of a 10-year programme, commenced in 1994 that laid the basis for application to human stem cells [Hearn, 2001]. As often in the past, the success for the human research ignored the essential milestones of earlier work in primate and other animal models.

In this paper, I explore the future of cloning and stem cell biology, arguing that rapid progress in this new and exciting field will be enhanced only if rigorous scientific method and the use of homologous animal research systems are employed. There are complex issues of ethics, risk and regulation that are highly likely to require a foundation of primate research with experimental systems from the marmoset, rhesus and perhaps the baboon over the next 10 years if stem cell biology is to deliver on its promise.

In making this argument I am not advocating a sequential approach. It is not necessary, desirable or possible for all steps to be clarified in the non-human primate before proceeding to the human. A strict sequential approach would have slowed the outcomes in many areas of biomedicine, including those of assisted reproductive technologies, aids immunology and vaccine development, parasitic diseases and pharmacology. The research should proceed in parallel, depending on the specific questions being asked, in primates and in the human. Based on the current state of knowledge of stem cell biology and its future development, we can now consider a few examples to show how scientific, ethical and regulatory requirements would be strengthened through cognate primate research.

Where to Draw the Line?

There is a set of issues in cloning and stem cell biology that can be arranged in increasing order of invasiveness and of ethical complexity. Assessment of these issues is further complicated through the dilemmas, choices and balance required in considering scientific rigour, research opportunity, ethical aspects of different religious or social views and the need for regulation or legislation. The approaches taken to these matters around the world in the past 5 years have differed widely.

Experimental Primate Research

The research protocols and procedures, ethics and regulation of non-human primate research is not the primary focus of this paper. This issue has been debated exhaustively over many decades and the stringent regulation of primate research in the USA, UK, Australia and other countries is well documented. The species suggested for research related to stem cell biology and cloning are not threatened or endangered. The focus of current experimental research is through the non-surgical recovery of embryos, so that the invasiveness of procedures is relatively slight. In future, should research expand to the realm of cell transfer experiments, and the adaptability and the applicability of therapeutic cloning to whole animals, a greater degree of invasiveness may be required. In this emerging arena of research, the need for primate and other mammalian models is likely to be essential for defined questions in cell biology and physiology that are now hard to predict. It is likely that current experimental and regulatory protocols would cover these new research directions.

Adult Stem Cells

As currently envisaged, cell therapies based on adult stem cells are likely to raise fewer ethical concerns than embryonic stem cells. There are now a significant number of publications claiming transdifferentiation of cells, for example from muscle to neuron or from blood precursor to heart. Whether these changes are functional and sustainable is still in question, since recent papers indicate that the transformation may be only temporary. Further work is necessary on the identification, isolation and characterisation of adult stem cells, since there are still many tissues and organs where this has not been achieved. It is not possible to predict whether study of adult stem cells will provide new cell therapies without a parallel study of embryonic stem cells. Currently, it is prudent to proceed in parallel. The field is still at its very early stages and further advances could easily change the appreciation and approach to stem cell biology. There do not appear to be serious ethical concerns about the derivation and study of adult stem cells, although the processing of transdifferentiation through an embryo stage would raise similar ethical issues to embryonic stem cells.

In vitro Fertilisation

It is often forgotten today that artificial reproductive technologies were opposed strongly at their inception in the 1980s by many people and also by research funding agencies in the UK and USA. Louise Brown was the first of one million babies born to date through IVF and related technologies, yet there is still a need to further understand the long term risks of IVF and procedures such as intracellular sperm injection (ICSI), especially when they overcome blocks to fertility in individuals who have inherent genetic or physiological obstacles to reproduction or who are at a stage of life when both sperm and eggs are ageing. In this arena, as in others noted below, technology on its own may facilitate objectives but without the ethical, social, economic and environmental considerations – while preserving informed individual choice – there can be a risk of fad outstripping fairness. Clearly, over the years, the perception of risk and ethical concern about IVF has diminished. Will the same be true for cloning and stem cell biology?

Frozen Embryos

'Excess' embryos from IVF procedures, stored for 5 years in freezers, must be discarded or 'allowed to expire' if they are not to be used in embryo transfer to the original parent or through donation. If these embryos are discarded, several countries now permit cells from the inner cell mass to be removed, dispersed and the embryonic stem cells isolated and grown for research purposes. The utilitarian argument of benefit – through advancing knowledge of cell differentiation, plasticity and perhaps later application – has usually won over the strict ethical consideration of the human embryo being sacrosanct in all circumstances. Embryos are made available under strict regulatory conditions in the UK, Australia and elsewhere. In the USA, embryos cannot be studied or stem cells derived from them, but this applies only to research funded by the Federal Government. In the private sector there are no such barriers – leaving a mixed signal of dual ethical standards.

Cloned Embryos

Embryos are cloned naturally in the formation of twins, artificially through embryo splitting and transfer and now also through the Dolly procedure where an adult cell is fused with an enucleated egg to produce a blastocyst from which embryonic stem cells can be derived. Once a blastocyst is formed – a result still only achieved in a very few percent of attempts – then it may be transferred for reproductive cloning as with Dolly or manipulated to produce stem cells for therapeutic cloning. These embryos and cells will carry the same defects as those of the initial fused cell and its transferred nucleus. Our knowledge is still rudimentary in this area, with enormous gaps in understanding how the enucleated egg can reprogramme the adult nucleus and broaden cell lineage choices to pluripotency. The ethical and regulatory management of this issue varies around the world. The UK, France, Singapore and Israel permit the formation of embryos for research – either by IVF or by cloning. Australia does not allow the cloning of embryos but is likely to review the position in a year. In the USA, cloning of embryos is prohibited in the public sector and permitted in the private sector. Attempts to ban embryo cloning in the United Nations are unlikely to achieve a consensus. Evidently there are significant variations in approach to this level of intervention and the position of such research is likely to evolve over the next few years.

Future Developments

The study of therapeutic cloning and of reproductive cloning is certain to expand greatly. The applications of these new technologies are only starting to be understood. As the potential applications increase, so will the need for homologous animal experimental systems where it is not possible or ethical to develop the field directly in the human. Examples are in the study of spinal cord or neuronal deficits and repair, pancreas and insulin replacements for the treatment of type 1 and type 2 diabetes, as well as the repair and reshaping of traumatic injury to muscle, liver and other tissues. There are fundamental opportunities as well in understanding the limits to the cell cycle and the choices in cell lineage as well as the agents that affect gene activation and expression and the down stream physiological effects. The tolerance limits of trans-differentiation between adult cell types have hardly been considered.

Reproductive cloning can be achieved naturally in identical twins and artificially through embryo splitting, as is routinely employed in animal production. Reproductive cloning through the Dolly procedure will certainly be used further in future, with research delivering a higher rate of success to birth and perhaps a lower rate of genetic damage or disorder. In some quarters there are calls for the technology to be applied to the conservation of endangered species and even to the recall of extinct species such as the woolly mammoth or the Tasmanian tiger. The present state of the art means that vast amounts of money would need to be invested in any such initiatives, raising the ethical and practical question whether any such funds should be spent in current conservation efforts to understand and accelerate breeding or to extend and manage existing environments. There is also a significant risk that those who use such 'visions' for fund raising are seriously damaging the credibility of science and operating under false pretences. Improbable high technology solutions should not be dressed up as practical conservation.

Conclusions

The new and emerging fields of mammalian cloning and human stem cell biology is only 5 years old, with as yet a rudimentary understanding of the potential or the realities of application. There are many scientific, ethical and regulatory issues that must be considered in advancing knowledge. The opportunities in agriculture, biomedicine and conservation are enormous. The timeframe is not likely to be short and there is a real risk of overstating the benefits. A balanced approach will need to analyse carefully the need for both non-human primate and human research appropriate to the scientific question and the ethical considerations. Primate model systems, based on knowledge of reproductive physiology, cell biology and immunology in species such as the marmoset, rhesus and baboon, are likely to be valuable alongside research in the human primate if the benefits are to be maximised.

Acknowledgements

I thank the many colleagues who worked with me in various aspects of our 10-year research programme to isolate embryonic stem cells from the marmoset and rhesus monkey [Hearn, 2001]. I am grateful for support from the MRC (UK), the NIH-NCRR (USA), the World Health Organization Department of Reproductive Health Research and the Universities of London, Wisconsin and the Australian National University. I thank my colleagues and friends in Kenya at the Institute of Primate Research and the departments of Physiology, Obstetrics and Gynaecology of the University.

References

Hearn JP: Embryo implantation and embryonic stem cell development in primates. Reprod Fertil Dev 2001;13:517–522.

Rhind SM, Taylor JE, De Sousa PA, King TJ, McGarry M, Wilmut I: Human cloning: Can it be made safe? Nature Reviews/Genetics 2003;4:855–864.

Thomson JA, Kalishman J, Hearn JP: Isolation of a primate embryonic stem cell line. Proc Natl Acad Sci USA 1995;92:7844–7848.

Thomson JA, Kalishman J, Golos TG, Durning M, Harris CP, Hearn JP: Pluripotent cell lines derived from common marmoset *(Callithrix jacchus)* blastocysts. Biol Reprod 1996;55:254–259.

Thomson JA, Itskovitz-Eldor J, Shapiro SS, Waknitz MA, Swiergiel JJ, Marshall VS, Jones JM: Embryonic stem cell lines derived from human blastocysts. Science 1998;282:1145–1147.

12

Embryonic-Endometrial Interactions at Implantation in Humans

C. Simon[a, b], *F. Dominguez*[a, b]

[a] Department of Gynaecology, Obstetrics and Pediatrics, School of Medicine, University of Valencia, and
[b] IVI Foundation, Valencia, Spain

It is known that successful implantation requires a functionally normal embryo at the blastocyst stage and a receptive endometrium, while a communication link between the two is also vital. This process is a highly regulated mechanism and numerous molecules take part in the aforementioned cross-talk. Here we present updated information, obtained in humans, concerning the molecules implicated in endometrial receptivity and the embryonic regulation of endometrial epithelial molecules such as chemokines.

Scientific knowledge of the endometrial receptivity process has been fundamental for the understanding of human reproduction, but so far none of the proposed biochemical markers for endometrial receptivity have been proven to be clinically useful. In this study we present new strategies based on molecular biology techniques that aim to clarify the fragmented information which exists in this field. We have done this using quantitative PCR and cDNA microarray analysis of endometrial epithelial-derived cell lines and endometrial samples so as to investigate the hierarchy of molecules at the mRNA level that are implicated in endometrial receptivity.

We analyzed differentially, using a combined approach of human endometrium and endometrial cell lines, the gene expression pattern of 375 human cytokines, chemokines and related factors, plus that of their receptors, in endometrial receptivity. We have compared said pattern in receptive (LH+7) versus pre-receptive (LH+2) human endometria and contrasted the results with that seen in the highly adhesive (to JAR cells and mouse blastocysts) cell line RL95-2 versus HEC-1A, a markedly less adhesive cell line. The human endometrial RL95-2 is an epithelial cell line derived from a moderately differentiated endometrial adenocarcinoma [1], with specific morphological and biological characteristics [2]. This cell line exhibits more pronounced adhesiveness for trophoblast-derived cells (JAR cells) [3] and mouse blastocysts [4] than any other human EEC line, including HEC-1-A and primary epithelium. The HEC-1-A cell line, in contrast, has poor adhesive properties, and exhibits a polarized distribution of integrins, while the RL95-2 cell line shows atypical features in adherens junctions, with non-polarized actin cytoskeleton and integrin distribution [5]. Embryonic adhesion experiments using mouse blastocysts showed pronounced receptive and non-receptive phenotypes in RL95-2 and HEC-1-A cells (81 vs. 46% of blastocyst adhesion, respectively), when compared to an intermediate adhesion rate in primary EEC cultured on extracellular matrix (67% of blastocyst adhesion) [4]. Therefore, we have used these cell lines as in vitro models for higher receptivity (RL95-2) and lower receptivity (HEC-1-A).

When we compared the differential gene expression pattern in receptive (LH+7) versus pre-receptive (LH+2), we identified genes that were already known to be expressed differently in the receptive and pre-receptive phases, such as PP14 (14.4-fold increase), osteopontin (3.7-fold increase), integrin $\alpha3$ (2.3-fold increase) and IL-1RtI (1.8-fold increase). However, we also detected a number of genes which had not been previously identified in the human endometrium and whose difference of expression in the pre-receptive (LH+2) and the receptive (LH+7) phases had not been described. These genes can be classified in different groups: extracellular matrix proteins (decorin), heparine-binding molecules (pleiotrophin), genes related to tyrosine kinases (EFNA2) and growth factors (BMP-7).

Other well-studied genes in the human endometrium are expressed in different change fold-increase in receptive endometrium. These include, among others, protease inhibitors such as TIMP1, -2 and -3, matrix metalloproteinase 13 (MMP-13) and insulin growth factor II (IGFII). Only four genes were minimally downregulated in the receptive endometrium, (between 1.32- and 1.89-fold increase). These genes were two interleukin receptors (IL-15R-α and IL-9R), bone morphogenetic protein 7 (BMP-7, a growth factor of the TGF-α family) and ephrin-A2 (a tyrosine kinase ligand).

We compared the results obtained after cDNA array hybridization of the endometrial cell lines, HEC-1-A versus RL95-2. The two highly expressed genes in the RL95-2 cell line were neurite growth-promoting factor 2 (NEGF2/midkine) and IGFBP-rP1, with a 16- and 12-fold change, respectively. The rest of the up-regulated genes were chemokines (GRO oncogene 1 and 2), growth factor receptors (erbB1 and TNFRSF16) and growth factors (TGF-α). The group of adhesion molecules was up-regulated (between 1.30- and 3.69-fold increase) and included EpCAM and integrins $\beta1$, $\alpha4$ and $\alpha1$. Remarkably, IGFBP-rP1 was the second most up-regulated gene in the two models investigated, showing a 4.6-fold increase in receptive versus pre-receptive endometria and an 11.8-fold increase in high-adhesive versus low-adhesive cell lines, respectively. RT-PCR was performed using the same RNA from the two cell lines and endometrial biopsies confirmed the RNA expression of the cDNA arrays. IGFBP-rP1 expression in both models exhibited a clear trend, echoing the results obtained in the cDNA arrays.

Of the remaining genes, only the matrix metalloproteinase 3 (TIMP-3) inhibitor and integrin $\beta4$ were up-regulated in both models. There were some discrepancies between these two models; up-regulated genes in the receptive endometrium, such as osteopontin and integrin $\alpha3$, were down-regulated in the high-adhesive cell line and down-regulated genes in the receptive endometrium, such as ephrin-A2, were up-regulated in the high-adhesive cell line.

In reproductive biology, chemokines have been implicated in crucial processes such as ovulation, menstruation, embryo implantation, parturition and pathological processes such as preterm delivery, HIV infection, endometriosis and ovarian hyperstimulation syndrome [6]. Accumulated evidence suggests that chemokines produced and received by the endometrial epithelium and the human blastocyst may be implicated in the molecular cross-talk during the implantation process.

Chemokine receptors belong to the superfamily of G-protein-coupled receptors (GPCRs). They have seven domains of 20–25 hydrophobic residues that form an α-helix and cross the plasma membrane, an extracellular N-terminus, three extracellular loops, three intracellular domains and an intracellular C-terminal tail. These receptors transmit information to the cell about the presence of chemokine gradients in the extracellular environment. They are named depending on the structure of their ligand (CXC or CC). CXCR4 is expressed on neutrophils, monocytes and B and T lymphocytes, and its primary ligand is the stromal derived factor-1 (SDF-1) [7]. CCR5, receptor for RANTES and macrophage inflammatory protein-1 (MIP-1) α and β is expressed on monocytes, dendritic cells, activated T lymphocytes and NK cells [8]. CCR2B, expressed in monocytes, basophils, dendritic cells, NK cells and activated T lymphocytes, is the main receptor for monocyte chemotactic proteins (MCP-1, -2, -3 and -4) [9]. CXCR1, a receptor for interleukin 8 (IL-8) and GCP-2, is expressed mainly in neutrophils and dendritic cells [10]. The binding of chemokines to their receptors is followed by the involvement of heterotrimeric G proteins [11] and the triggering of intracellular second messengers such as cAMP and calcium. An impressive consequence of the binding of chemokines to their receptors on leukocytes are the morphological changes which this provokes and by which the cytoskeleton is rearranged, integrin-mediated focal adhesions are formed and the cell binds and detaches from the substrate in a coordinated manner, with extension and retraction of pseudopods being responsible for directional migration [11].

A specific molecular crosstalk between embryo and endometrium has been reported during the human implantation process [12]. The endometrial epithelium is an important element, being both the site where the molecular interactions between the embryo and endometrium seem to be initiated [13,14] and the source and target of chemokines [15].

CXCR1 (IL-8 receptor), CCR2B (MCP-1 receptor) and CCR5 (RANTES receptor) mRNAs are highly regulated throughout the menstrual cycle, with maximal expression in the luteal phase. We found that mRNA expression for the three receptors showed a progesterone-dependent regulation starting in the early secretory phase, continuing throughout the mid secretory phase and peaking in the late secretory phase (612-fold increase for CCR5, 419-fold increase for CXCR1 and 657-fold increase for CCR2B). Unlike the other receptors, CXCR4 receptor mRNA presented a specific up-regulation in the mid luteal when compared to the early and late luteal phases (17.7-fold increase in mid luteal phase). This illustrates that this receptor is specifically up-regulated during the implantation window and that its lowest expression occurs in the early secretory phase. Furthermore, two of these receptors, CCR2B and CCR5, have been identified in the human blastocyst [16].

We have studied the four receptors at protein level, throughout the menstrual cycle. CXCR1 is at its peak in the early and mid secretory phases. Similarly, CXCR4 (SDF-1 receptor) displays high staining across the whole cycle, with maximal expression in the mid secretory phase. CCR receptors in general show lower expression in the endometrium than CXC receptors. CCR2B staining appears in the late proliferate phase and reaches a moderate signal in the early secretory phase, maintaining a low to moderate staining in the rest of the cycle. The CCR5 receptor signal starts in the late proliferate phase and remains low to moderate in different compartments.

We used confocal microscopy to observe chemokine receptors in cultured endometrial epithelial cells (EEC) and CXCR1, CXCR4 and CCR5 receptors showed a polarization when a human blastocyst was present. The embryonic impact on immunolocalization and polarization of chemokine receptors CXCR1, CXCR4, CCR5 and CCR2B in cultured EEC was investigated using our apposition model for human implantation. When the blastocyst was absent the first three receptors produced a barely detectable staining in only a few cells at the EEC monolayer. However, when a human blastocyst was present there was an increase in the number of stained cells for CXCR1, CXCR4, and CCR5 and polarization of these receptors in one of the cell poles of the endometrial epithelium. Immunolocalization and polarization changes in the CCR2B receptor were not observed in the EEC monolayer and this receptor was not up-regulated by the presence of the human blastocyst.

Therefore, we have detected the presence of the RANTES and MCP-1 receptors, CCR2 and CCR5 in the human blastocyst, and we have found an up-regulation of endometrial CXCR1 (419-fold increase), CCR5 (612-fold increase) and CCR2B mRNA (657-fold increase) during the luteal phase, which peaks in the premenstrual endometrium

An array of different chemokines are expressed [17] and produced in the human endometrium at the time of implantation [6, 18] and these chemokines could be immobilized by low-affinity binding to heparin-bearing proteoglycans on the vascular endothelial or epithelial surface, thereby facilitating the oligomerization of chemokines [19]. This effect permits effective presentation of chemokines to cells or groups of cells, which then respond. In this way, variations in the availability of these chemokines would affect the ability of a ligand to trigger homo- or heterodimerization. At low concentrations of chemokines, receptor heterodimerization is favoured and cell adhesion triggered [20]. Caballero-Campo et al. [15] have investigated the hormonal and embryonic regulation of IL-8, RANTES and MCP-1 in the endometrium. IL-8 and MCP-1 were present in the glandular and luminal epithelium, and RANTES was mainly localized on stromal cells. IL-8 and MCP-1 were up-regulated in the presence of E_2 and P. IL-8 mRNA expression and protein were up-regulated in the presence of human blastocyst using our co-culture model.

References

1 Way DL, Grosso DS, Davis JR, Surwit EA, Christian CD: Characterization of a new human endometrial carcinoma (RL95-2) established in tissue culture. In Vitro 1983;19:147–158.

2 Tinel H, Denker HW, Thie M: Calcium influx in human uterine epithelial RL95-2 cells triggers adhesiveness for trophoblast-like cells. Model studies on signalling events during embryo implantation. Mol Hum Reprod 2000; 6:1119–1130.

3 John NJ, Linke M, Denker HW: Quantitation of human choriocarcinoma spheroid attachment to uterine epithelial cell monolayers. In Vitro Cell Dev Biol Anim 1993;29:461–468.

4 Martin JC, Jasper MJ, Valbuena D, Meseguer M, Remohi J, Pellicer A, Simon C: Increased adhesiveness in cultured endometrial-derived cells is related to the absence of moesin expression. Biol Reprod 2000;63:1370–1376.

5 Thie M, Fuchs P, Butz S, Sieckmann F, Hoschutzky H, Kemler R, Denker HW: Adhesiveness of the apical surface of uterine epithelial cells: the role of junctional complex integrity. Eur J Cell Biol 1996;70:221–232.

6 Simón C, Caballero-Campo P, García-Velasco JA, Pellicer A: Potential implications of chemokines in reproductive function: an attractive idea. J Reprod Immunol 1998;38:169–193.

7 Nagasawa T, Nakajima T, Tachibana K, Iizasa H, Bleul CC, Yoshie O, Matsushima K, Yoshida N, Springer TA, Kishimoto T: Molecular cloning and characterization of a murine pre-B-cell growth-stimulating factor/stromal cell-derived factor 1 receptor, a murine homolog of the human immunodeficiency virus1 entry coreceptor fusin. Proc Natl Acad Sci USA 1996; 93:14726–14729.

8 Chantakru S, Kuziel WA, Maeda N, Croy BA: A study on the density and distribution of uterine natural killer cells at mid pregnancy in mice genetically-ablated for CCR2B, CCR 5 and the CCR5 receptor ligand, MIP-1 alpha. J Reprod Immunol 2001;49:33–47.
9 Polentarutti N, Allavena P, Bianchi G, Giardina G, Basile A, Sozzani S, Mantovani A, Introna M: IL-2-regulated expression of the monocyte chemotactic protein-1 receptor (CCR2B) in human NK cells: characterization of a predominant 3.4-kilobase transcript containing CCR2B and CCR2A sequences. J Immunol 1997;158:2689–2694.
10 Wuyts A, Proost P, Lenaerts JP, Ben-Baruch A, Van Damme J, Wang JM: Differential usage of the CXC chemokine receptors 1 and 2 by interleukin-8, granulocyte chemotactic protein-2 and epithelial-cell-derived neutrophil attractant-78. Eur J Biochem 1998;255:67–73.
11 Ward SG, Bacon K, Westwick J: Chemokines and T lymphocytes: more than an attraction. Immunity 1998;9:1–11.
12 De los Santos MJ, Mercader A, Francés A, Portoles E, Remohi J, Pellicer A, Simon C: Role of endometrial factors in regulating secretion of components of the immunoreactive human embryonic interleukin-1 system during embryonic development. Biol Reprod 1996;54:563–574.
13 Galan A, O'Connor JE, Valbuena D, Herrer R, Remohi J, Pampfer S, Pellicer A, Simon C: The human blastocyst regulates endommtrial epithelial apoptosis in embryonic adhesion. Biol Reprod 2000;63:430–439.
14 Meseguer M, Aplin D, Caballero-Campo P, O'Connor JE, Martin JC, Remohi J, Pellicer A, Simon, C: Human endometrial mucin MUC1 is up-regulated by progesterone and down regulated in vitro by the human blastocyst. Biol Reprod 2001;64:590–601.
15 Caballero-Campo P, Dominguez F, Coloma J, Meseguer M, Remohi J, Pellicer A, Simon C: Hormonal and embryonic regulation of chemokines IL-8, MCP-1 and RANTES in human endometrium during the window of implantation. Mol Hum Reprod 2002;8:375–384.
16 Dominguez F, Galan A, Martin JJ, Remohi J, Pellicer A, Simón C: Hormonal and embryonic regulation of chemokine receptors CXCR1, CXCR4, CCR5 and CCR2B in the human endometrium and the human blastocyst. Mol Hum Reprod 2003;9:189–198.
17 Kao LC, Tulac S, Lobo Imani B, Yang JP, Germeyer A, Osteen K, Taylor RN, Lessey BA, Giudice LC: Global gene profiling in human endometrium during the window of implantation. Endocrinology 2002;143:2119–2118.
18 Kayisli UA, Mahutte NG, Arici A: Uterine chemokines in reproductive physiology and pathology. Am J Reprod Immunol 2002;47:213–221.
19 Hoogewerf AJ, Kuschert GSV, Proudfoot AE, Chung CW, Cooke RM, Hubbard RE, Wells TN, Sanderson PN: Glycosamins mediate cell surface oligomerization of chemokines. Biochem 1997;36:13570–13578.
20 Mellado M, Rodriguez-Frade JM, Vila-Coro AJ, Fernandez S, Martin de Ana A, Jones DR, Toran JL, Martinez-A C: Chemokine receptor homo- or heterodimerization activates distinct signaling pathways. EMBO J 2001; 20:2497–2507.

13

Implantation in the Baboon

A.T. Fazleabas, Z. Strakova, J.J. Kim

Department of Obstetrics and Gynecology, University of Illinois at Chicago, Chicago, IL, USA

Introduction

Implantation is a complex spatio-temporal interaction between the genotypically different embryo and the mother. Success of this event requires the synchronization of development and effective biochemical communications from both sides. Implantation can be divided into three distinct phases: The first phase, regulated by estrogen and progesterone, is characterized primarily by changes in both the luminal and glandular epithelial cells in preparation for blastocyst apposition and attachment. The second phase is the further modulation of these steroid induced changes in both epithelial and stromal cells by embryonic signals. The final phase is associated with trophoblast invasion and the remodeling of the endometrial stromal compartment, a process known as decidualization.

Chorionic gonadotrophin (CG), which is a major embryonic signal in the primate, is a glycoprotein hormone synthesized and secreted by the trophoblast. Various isoforms exist in plasma, urine and blastocyst culture medium which are a result of post translational modifications. The exponential secretion of CG and its long circulatory half-life extends the life span of corpus luteum to maintain the supply of progesterone during the first 6–8 weeks of pregnancy. To study the direct effects of CG in the uterus, we used the baboon *(Papio anubis)* as a non-human primate model. In vivo stimulation with CG during window of uterine receptivity results in further morphological and biochemical modifications of the receptive endometrium.

CG Modulation of Epithelial Cells in the Endometrium of the Baboon

Luminal Epithelium. When CG was infused into the uterine cavity of cycling baboon from day 6 to day 10 post ovulation, the characteristic epithelial plaque reaction was observed in the luminal epithelium. Plaque reaction is a morphological transformation of epithelial cells, which is characterized by the hypertrophy, hyperplasia and differentiation of the surface epithelium and cells in the neck glands that round up and form a multicellular pad. The induction of the plaque reaction is probably activated by the infiltration of the inflammatory cells into the uterus. It may be a prerequisite for the initial phase of blastocyst-endometrial interactions. Other factors from the ovary are required for this reaction since it is not observed in the ovariectomized baboons. This reaction is partially inhibited by a progesterone receptor antagonist ZK137.316 (PRa), which also induces the re-expression of progesterone receptor (PR) and estrogen receptor α (ERα) in the luminal epithelial cells. This reaction has not observed in human.

Glandular Epithelium. CG induces morphological changes in endometrial glandular structures that resemble those observed in the pregnant baboon of the same gestational age. Glands in the functionalis and basalis layers are distended and convoluted. The glandular secretory function is also increased and corresponds to structural changes. Glycodelin expression and production is increased by CG. Glycodelin is the major secretory protein of the endometrium during secretory phase and pregnancy. The synthesis of glycodelin parallels the rise and decline of CG in the peripheral circulation. Glycodelin is associated with immunosuppression, inhibition of sperm-egg binding and induction of epithelial cell differentiation. The functions of glycodelin during implantation are under investigation. Glycodylin production is inhibited by PRa which also induces the re-expression of PR and ERα in the glandular epithelium.

CG Modulation of the Stromal Cells in the Endometrium of the Baboon

CG induced the characteristic stromal edema due to increased vascular permeability and congestion. In addition, stromal fibroblast cells in the subepithelial region showed a differentiation in response to CG by the immunostaining for α-smooth muscle actin (α-SMA). We propose that the induction of α-SMA in the stromal fibroblasts occurs in response to integrins on the stromal cell membranes binding to secreted extracellular matrix (ECM) proteins. The specific changes in integrin expression and ECM secretion are also observed in baboon endometrium in response to CG. These 'out side in' signals induce changes in the actin cytoskeleton and generates the signals for cellular responses that are essential for stromal cell proliferation and differentiation during the initial stage of pregnancy. However, decidualization in baboon requires other embryonic signals in contrast to human. The expression of α-SMA in stromal fibroblasts is completely inhibited by PRa that also induces the re-expression of ERα in the stromal cells.

In vitro data also support the effects of CG on the stromal fibroblasts. Our studies using baboon stromal cells suggest that CG can promote decidualization since the addition of CG to steroid-treated cells inhibits apoptosis and enhances differentiation as determined by IGFBP-1 expression, a marker of differentiation. In a manner similar to endometrial cells, differentiating stromal cells show an increase in COX-2 transcription paralleled by an increase with PGE_2. Studies with human stromal fibroblasts also substantiate this observation that CG can promote decidualization. CG in the presence estrogen and progesterone can induce a functional differentiation as determined by an increase in the transcription of prolactin, a gene that is a marker of decidualization in the stromal cells.

In conclusion, these data demonstrated that CG has physiological effects on the primate uterus during the period of uterine receptivity. They also support the concept that the blastocyst plays an active role in transforming the epithelial barrier into a localized gateway to the stromal compartment. Our in vivo results also showed that PRa could completely or partially inhibit these changes. This leads to the paradox that the high level of steroids especially progesterone is required for uterine receptivity, implantation and pregnancy but their direct actions (through PR or ERα) inhibit the cellular and molecular changes induced by the embryonic signals in the uterus.

Decidualization in the Baboon

Decidualization, which involves the transformation of stromal fibroblasts to decidual cells, is the major change that occurs in the primate endometrium after conception. In primates and rodents, the uterine endometrial stromal cells differentiate to decidual cells following the establishment of pregnancy. Decidual cells play an important role in implantation and provide nutritional support for embryo. Decidual cells are also believed to produce factors that control trophoblast invasion and protect the embryo from maternal immune rejection. During the process of decidualization in the primate, fibroblast-like stromal cells change morphologically into polygonal cells and begin to express specific decidual proteins. This is manifested by the downregulation of α-SMA expression and the induction of insulin-like growth factor binding protein-1 (IGFBP-1). Previous studies in the baboon have clearly demonstrated that IGFBP-1 gene expression in the endometrium is a conceptus-mediated response. Subsequent studies in vitro established that IGFBP-1 gene expression in decidualizing stromal fibroblasts requires the presence of both hormones and cAMP. This induction is associated with a concomitant decrease of α-SMA expression in vivo and in vitro. Since IL-1β is expressed both in the progestational endometrium and in trophoblast cells we evaluated IL-1β as one possible factor that could influence differentiation of stromal cells into decidual cells. IL-1β has been reported to be actively involved in fetal-maternal interactions, but its role in decidualization has not been clarified. In addition, IL-1 can modulate changes in the cytoskeleton and induce cyclooxygenase-2 (COX-2) gene expression. Our data would suggest that IL-1β activates a signaling pathway that induces COX-2 expression. COX-2 in turn increases PGE_2 which can increase intracellular cAMP via activation of the EP2 and EP4 receptor. The cAMP in synergism with progesterone results in the induction of IGFBP-1 expression. Induction of IGFBP-1 in these cells is transcriptionally regulated by FKHR and HOXA10 which together activate the IRE on the IGFBP-1 promoter. Coincident with the induction of COX-2, IL-1β also induces metalloproteinase-3 (MMP-3) expression in stromal fibroblasts. We hypothesize that the local action of MMP-3 dissociates the surrounding extracellular matrix resulting in the loss of focal adhesion complexes and the alteration in the actin cytoskeleton. This dissociation is the necessary pre-requisite for decidualization and IGFBP-1 induction. In summary, our studies have demonstrated that a coordinated sequence of events that require a signal from the conceptus are necessary to induce stromal cell differentiation and decidualization. We suggest that these changes are critical to ensure prolonged maintenance of endometrial function during gestation and facilitate trophoblast invasion.

Thus, in the baboon implantation is characterized by a dynamic process that requires an intimate communication between the implanting embryo and the maternal endometrium. A fundamental question that remains to be resolved is what the functional significance of the gene products are? Many of them are expressed both in the human and baboon. The availability of the baboon model will enable us to undertake in vivo studies to address some of the questions as they pertain to the physiological importance of the hormone- and conceptus-induced maternal gene products.

14

Morphology, Endocrinology and Paracrinology of Embryo Growth and Implantation in the Rhesus Monkey

J. Sengupta, D. Ghosh

Department of Physiology, All India Institute of Medical Sciences, New Delhi, India

The marked similarities in the endocrinology, and the intrusive nature of blastocyst implantation between the human and the rhesus monkey makes the latter a useful experimental primate model to study the early events in gestation and for testing innovative strategies for post-coital contraception.

Embryo

We have earlier reported based on 304 menstrual cycles (241 were ovulatory cycles) studied in outbred, proven-fertile rhesus monkeys that 62% of mated, ovulatory cycles yielded age- and stage-matched preimplantation embryos from the reproductive tract on its surgical flushing [1]. The transport of egg through the oviduct takes place during days 3 to 4 after ovulation in the human [2]. In the rhesus monkey, pre-implantation embryos at 12- to 16-cell stages or early morula stage enter the uterus by 72 h after ovulation [1]. Morphological and ultrastructural studies of the Fallopian tube and the endometrium in the human and the rhesus monkey have revealed cyclical changes during the period of egg transport through the oviduct reflecting stage-dependent functional differentiation along with characteristic distribution of receptors for estrogen and progesterone [3]. Evidence that the Fallopian tube secretes growth factors, and that their receptors are localized in early stage human preimplantation embryos supports the hypothesis that tubal secretions are embryotropic for early embryo development [4]. Schramm and Bavister [5] reported that macaque embryos can form morulae in a chemically defined, protein-free medium, however, blastocyst formation and its hatching required some serum components. Similarly co-culture studies using sequential human oviduct and endometrial cells were shown to favour blastocyst formation [6]. In studies designed to test whether progesterone is necessary for promoting early embryo development, there is evidence to suggest that progesterone-dependent milieu of the reproductive tract is essential to promote morula to blastocyst transition and embryo viability in the rhesus monkey [7, 8].

Ultrastructural study of pre-implantation embryos of human and non-human primates reveal striking similarities in their developmental patterns based on studies performed using embryos recovered from in situ flushing and in vitro fertilization. Cleavage stage embryos from rhesus monkeys show large blastomeres often of unequal size, with large inter-blastomeric spaces frequently containing cellular debris, while they generally lack typical junctional complexes and cell surface microvilli [9]. In early cleavage stage embryos, mitochondria are structurally undifferentiated and generate ATP at relatively low levels compared with that in moruale and blastocyst stage embryos. Maturation of mitochondria from large convoluted electron-dense structures to short, elongated, electron-lucent forms with distinctive lamellar cristae is a feature of maturation of this cellular organelle in rhesus monkey and baboon morulae [8, 9]. It has been reported that the delamellated state of the cristae leads to inactivation of oxidative phosphorylation to protect the early embryo from toxic respiratory end products [10]. In human and non-human primates, morula to blastocyst transition is accompanied by compaction of blastomeres with development of apical junctional complexes in the form of tight junctions, gap junctions and desmosomes. In the morula stage, closely parallel cell membranes often showed presence of caveolae and tight junctional complexes with intermediate filament investment [8, 9, 11].

An interesting feature of blastocyst stage lies in the differentiation of inner cell mass, formation of endoderm, and cellular distinctions of embryonic and abembryonic trophoblast cells. Patches of basal laminae were observed to be present beneath mural trophoblast with underlying ICM and endoderm cells. Polar trophoblast cells were flattened, squamous shaped, dense with apical regions showing numerous vacuoles, vesicles, microvilli, and extensive well-developed junctional complexes between neighbouring cells, cells of ICM, and endodermal cells. Azonal blastocysts have regions of large, irregular surface projections of syncytial trophoblast cells just peripheral to ICM suggesting that such areas may have been adherent to uterus when blastocysts were flushed from the uterus [9, 12]. In rhesus monkey blastocysts, budding virus-like particles were distinguished on nuclear envelope, within cytoplasm and were found budding from microvillous projections of mural trophoblast cells. Type C retroviruses have been reported in germ cells, in cleavage stages and in syncytial trophoblast cells of human and baboon placenta and in syncytial trophoblast cells of in vitro fertilized rhesus ova [13]. It has been suggested that virus particles detected at the time of implantation in marmosets may contribute to the formation of syncytium [14].

While limitations exist in examining embryo morphology commensurate with its potential for implantation, we conducted a parallel study of morphology and viability for implantation in the preimplantation stage embryos of the rhesus monkey. Marked changes were noted in trophoblast cells with absence of typical junctional complexes and differentiated features of apical and basal surfaces in non-viable blastocysts [8]. In a comparative study of bovine blastocysts developed in vivo and in vitro, a similar lack of well-developed junctional complexes between trophoblast cells and less well-developed microvilli was observed suggestive of physiological inadequacies and reduced viability [15, 16].

Receptive Endometrium

The physiological and biochemical determinants for endometrial receptivity towards embryo attachment and implantation remain poorly understood in the human. The endocrine correlates of endometrial receptivity and implantation are now documented. Progesterone has been shown to induce a basic drive and an innate releasing mechanism towards implantation involving both embryo and endometrium; however, once it is initiated the process may proceed through certain steps in a fixed action pattern so that even vacuum stimuli may induce to some extent a mimicking response [17]. Despite the fact that serum concentration of progesterone is highest during the mid-luteal phase of an ovulatory cycle, it appears possible that endometrial maturity towards receptivity does not require high concentration of progesterone in peripheral circulation, and that implantation stage embryos can withstand a partial lack of progesterone for a limited period of time [18]. In the macaque, a differential pattern of progesterone receptor is observed in gland, stroma and vascular cells of oviduct and endometrium during 'receptive window' of implantation, blockade of progesterone action leads to glandular and vascular dysfunctions and associated non-receptivity for blastocyst implantation [19].

Several endometrial factors and functions which are regulated by luteal phase progesterone appear as robust associates of endometrial receptivity and ovo-implantation. Using high-density oligonucleotide microarray technology, Kao et al. [20] have shown that there is specific up- and down-regulation of genes in human endometrium during the 'window of implantation' compared with late proliferative stage. However, we believe that it will be important to perform studies in endometrium from proven conception cycle with an existing 'dialogue' between embryo and endometrium in order to define the expression of the cluster of genes in maternal endometrium associated with preimplantation embryo development and implantation [21].

Implantation

Anthropoid blastocysts generally attach to uterine surface epithelium at embryonic pole, however, fine structural and molecular biological studies of blastocyst apposition and adhesion to epithelium have not been undertaken. A role of matrix metalloproteinases (MMPs) in the process of syncytial trophoblast intrusion between epithelial cells may be suggested (unpublished observation), and that trophoblastic vesicles grown in vitro using marmoset blastocysts are similarly capable of MMP-2 proenzyme secretion [22]. Trophoblast cells derived from rhesus blastocysts may prove useful for studies on epithelial invasion and formation of trophoblast plate for in vitro implantation.

Following trophoblast invasion, three characteristic endometrial responses appear: a vascular reaction characterized by edema, vascular congestion and endothelial cell hypertrophy, glandular hyperplasia and decidual transformation of stromal cells with associated influx of large endometrial granulated lymphocytes [23, 24]. In the rhesus monkey and the baboon, an additional response of epithelial plaque cells are found at the implantation site. Similar endometrial responses to blastocyst implantation were observed in the early implantation stage samples from women, except plaque cells were not reported [25]. An endocrine-paracrine network involving progesterone and growth factors has been shown to operate in regulating trophoblast cell proliferation and their migration, stromal and vascular cell remodeling and glandular functions in the implantation site of the rhesus monkey collected from day 12 of gestation (blastocyst implantation begins on estimated day 9) [26–28]. We hypothesize that a complex interaction of a large family of factors at different levels of hierarchies of the system provides redundancy as well as homeodynamic stability to the process of blastocyst implantation under a central controller action of progesterone.

Supported by the WHO-Rockefeller Foundation and the CONRAD-CICCR.

References

1 Ghosh D, Sengupta J: Patterns of ovulation, conception and pre-implantation embryo development during the breeding season in rhesus monkeys kept under semi-natural conditions. J Endocrinol 1992;127:168–173.
2 Croxatto HR, Diaz D, Fuentealba B, et al: Studies on the duration of egg transport in the human oviduct. I. The time interval between ovulation and egg recovery from the uterus in normal women. Fertil Steril 1972;23:447–458.
3 Brenner RM, Maslar IA: The primate oviduct and endometrium; in Knobil E, Neill JD (eds): The Physiology of Reproduction. New York, Raven Press, vol 1, pp 303–330.
4 Smotrich DB, Stillman RJ, Widra EA, et al: Immunocytochemical localization of growth factors and their receptors in human pre-embryos and Fallopian tubes. Hum Reprod 1996;11:184–190.
5 Schramm RD, Bavister BD: Development of in vitro fertilized primate embryos into blastocysts in a chemically defined, protein-free culture medium. Hum Reprod 1996;11:1690–1697.
6 Bongso A, Fong C-Y, Ng S-C, Ratnam S: Human embryonic behaviour in a sequential human oviduct-endometrial coculture system. Fertil Steril 1994; 61:976–978.
7 Ghosh D, Kumar PGLL, Sengupta J: Early luteal phase administration of mifepristone inhibits preimplantation embryo development and viability in the rhesus monkey. Hum Reprod 1997;12:575–582.
8 Ghosh D, Lalitkumar PGL, Wong V, et al: Preimplantation embryo morphology following early luteal phase anti-nidatory treatment with mifepristine (RU486) in the rhesus monkey. Hum Reprod 2000;15:180–188.
9 Enders AC, Schlafke S: Differentiation of blastocyst in the rhesus monkey. Am J Anat 1981;162:1–21.
10 Shepard TH, Muffley LA, Smith LT: Mitochondrial ultrastructure in embryos after implantation. Hum Reprod 2000;15:218–228.
11 Lopata A, Kohlman D, Johnston I: The fine structure of normal and abnormal human embryos in culture; in Beier HM, Lindner HR (eds): Fertilization of the human egg in vitro. Berlin, Springer pp 189–210.
12 Sengupta J, Ghosh D: Blastocyst-endometrium interaction at impantation in the rhesus monkey. J Reprod Immunol 2002;53:227–239.
13 Enders AC, Boatman D, Morgan P, Bavister BD: Differentiation of blastocysts derived from in vitro fertilized rhesus monkey ova. Biol Reprod 1989; 41:715–727.
14 Smith CA, Moore HDM: Expression of C-type viral particles at implantation in the marmoset monkey. Hum Reprod 1988;3:95–398.
15 Abe H, Otoi T, Tachikawa S, et al: Fine structure of bovine morulae and blastocysts in vivo and in vitro. Anat Embryol 1999;199:519–527.
16 Fleming TP, Ghassemifar MR, Sheth B: Junctional complexes in the early mammalian embryo. Semin Reprod Med 2000;18:185–193.
17 Ghosh D, Sengupta J: Endometrial receptivity for implantation. Another look at the issue of peri-implantation oestrogen. Hum Reprod 1995;10: 1–2.
18 Edwards RG Implantation, interception and contraception. Hum Reprod 1994;9:73–87.
19 Ghosh D, Sengupta J: Recent developments in endocrinology and paracrinology of blastocyst implantation in the primate. Hum Reprod Update 1998;4:153–168.
20 Kao LC, Tulac S, Lobo S, et al: Global gene profiling in human endometrium during the window of implantation. Endocrinology 2001;143:2119–2138.
21 Ghosh D, Roy A, Sengupta J, Johannisson E: Morphological characteristics of preimplantation stage endometrium in the rhesus monkey. Hum Reprod 1993;8:1579–1587.
22 Franek AD, Salamonsen LA, Lopata A: Marmoset monkey trophoblastic tissue growth and matrix metalloproteinease secretion in culture. J Reprod Fertil 1999;117:107–114.
23 Enders AC, Welsh AO, Schlafke S: Implantation in the rhesus monkey: endometrial responses. Am J Anat 1985;173:147–169.
24 Sengupta J, De P, Ghosh D Implantation of the primate embryo. Curr Sci 1995;68:363–373.
25 Hertig AT, Rock J, Adams EC: A description of 34 human ova within the first 17 days of development. Am J Anat 1956;98:435–494.
26 Ghosh D, Dhara S, Kumar A, Sengupta: J Immunohistochemical localization of receptors for progesterone and oestradiol-17β in the implantation site of the rhesus monkey. Hum Reprod 1999;14:505–514.
27 Ghosh D, Sharkey A, Charnock-Jones DS, et al: Expression of vascular endothelial growth factor (VEGF) and placental growth factor (PlGF) in conceptus and endometrium during implantation in the rhesus monkey. Mol Hum Reprod 2000;6:935–941.
28 Dhara S, Lalitkumar PGLL, Sengupta J, Ghosh D: Immunohistochemical localization of insulin-like growth factors I and II at the primary implantation site in the rhesus monkey. Mol Hum Reprod 2001;7:365–371.

15

The Olive Baboon *(Papio anubis)*: A Potential Animal Model to Study the Function of Human Leukocyte Antigen-G (HLA-G)

D.K. Langat[a,c], P.J. Morales[a], A.T. Fazleabas[b], J.M. Mwenda[c], J.S. Hunt[a]

[a] Department of Anatomy and Cell Biology, University of Kansas Medical Center, Kansas City, Kansas, and [b] Department of Obstetrics and Gynecology, University of Illinois at Chicago, Chicago, IL, USA; [c] Institute of Primate Research, Karen, Nairobi, Kenya

Abstract

The human class Ib major histocompatibility complex (MHC) molecule, *HLA-G*, is unique in its limited polymorphism, high expression in the placenta and generation of multiple transcripts by alternative splicing. The proteins encoded by these transcripts are believed to modulate maternal-fetal immunological relationships during pregnancy. The baboon placenta expresses *Paan-AG*, a novel MHC molecule that is evolutionarily related to the *MHC-A* locus but shares unique characteristics with *HLA-G*. In this brief review, we present evidence suggesting that *Paan-AG* may be the functional homologue of *HLA-G*, and propose that the baboon would comprise an excellent animal model for functional studies of HLA-G proteins in human pregnancy.

Introduction

The mammalian fetus has been described as a semi-allograft because the father contributes half of the genes, and expression of the paternal genes by the fetus would be expected to result in rejection of the fetus by the maternal immune system as is the case in allograft rejection. However, this is not known to occur, and pregnancy survives to term more often than not. The potential mechanisms by which the fetus evades the hostile maternal immune system have

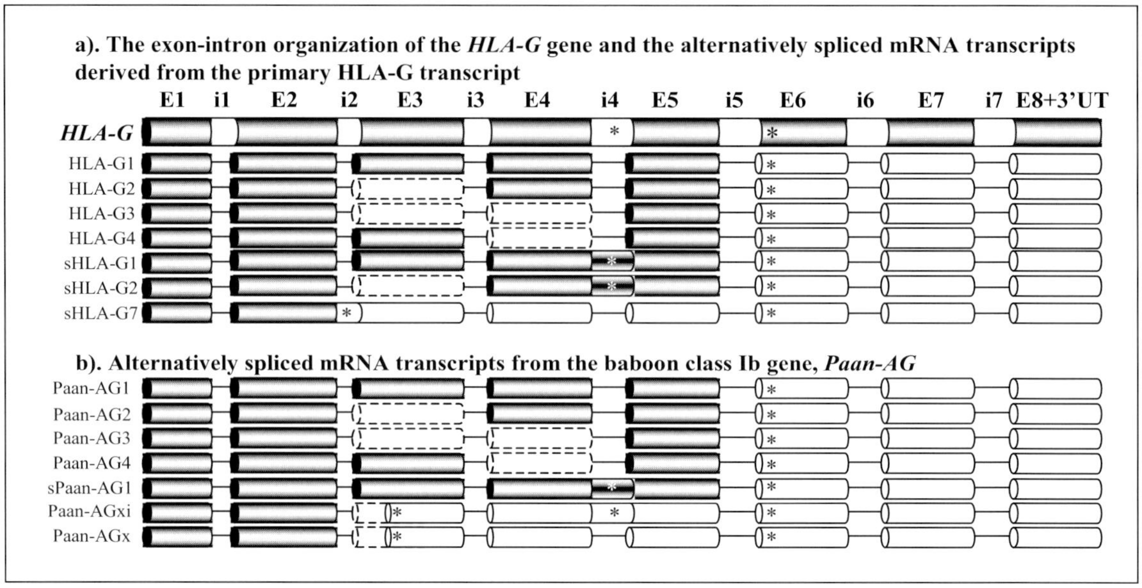

Fig. 1. Comparison of the alternatively spliced transcripts derived from *HLA-G* and *Paan-AG* genes. The top panel (**a**) shows the exon-intron organization of the *HLA-G* gene (HLA-6.0, Acc. No. J03027, [13]) and the HLA-G alternatively spliced transcripts that have been described so far in humans. The lower panel (**b**) shows the alternatively spliced transcripts of *Paan-AG* gene that were isolated from the baboon placenta. The black boxes represent the exons that are translated to proteins while the white ones represent the exons that are not translated due to the presence of a premature stop codon (*). The boxes with dashed borders represent the exons that are spliced out in each cDNA. E1–E8, exons 1–8; i1–i7, introns 1–8; 3′UT, 3′ untranslated region. Figure modified from ref. [8].

been described elsewhere [1–4]. One of these mechanisms is regulation of expression of the major histocompatibility complex antigens at the maternal-fetal interface. In humans, expression of MHC antigens at the maternal-fetal interface (mainly the placenta and extraplacental membranes) is strictly regulated [5]. The polymorphic class Ia MHC genes, human leukocyte antigen (*HLA*)-*A* and *HLA-B* as well as class II molecules are not expressed in the placenta. It is only the class Ia gene *HLA-C* and the class Ib genes *HLA-E* and *HLA-G* that are expressed. *HLA-G* has generated considerable interest in pregnancy immunology due to it potential role in the modulation of the maternal immune system during pregnancy and in other contexts.

Although there is substantial evidence supporting the idea that HLA-G plays an important role in modulating the maternal immune system during pregnancy [6], some of these reports are contradictory and the results have generated more questions than answers [7]. This is mainly because most of the studies on HLA-G are based either on in vitro experiments or observations on normal and pathological pregnancies [8]. Ethical concerns do not allow certain in vivo experiments to be performed in humans, thus an appropriate animal model would be useful. Non-human primates, being phylogenetically close to humans, offer suitable models for functional studies, and have been used extensively as models in biomedical research [9]. Studies in our laboratory have identified the olive baboon *(Papio anubis)* as a potential animal model for *HLA-G* functional studies, and in this brief review, we present evidence indicating that this animal would be a useful model.

Baboons Express a Potential Functional Homologue of HLA-G

The olive baboon *(Papio anubis)* is an Old World monkey found in Africa, which belongs to the family *Cercopithecidae*. The members of this family of non-human primates have homologues of both class Ia (*HLA-A* and *HLA-B*) and class Ib (*HLA-E* and *HLA-G*) genes. However, studies have shown that the *HLA-G* homologues (designated *MHC-G*) in the species that have been evaluated so far are inactivated [8, 10]. The *MHC-G* genes in these species contain stop codons in a restricted area of exon 3 [10], thus precluding translation. However, the olive baboon and the rhesus monkey express a novel gene (designated *Paan-AG* in the baboon and *Mamu-AG* in the rhesus monkey) in the placenta [11, 12]. The nucleotide and amino acid sequence analyses of these loci suggest that they are closely related to the *MHC-A* locus. These genes cluster with *MHC-A* genes on phylogenetic analysis and do not have the unique amino acids characteristic of HLA-G protein [8, 11, 13]. However, the genes have characteristics similar to *HLA-G* (described below), suggesting they may be functional homologues of *HLA-G*.

Structural Similarities between *HLA-G* and *Paan-AG*

The crystal structures of HLA-G and Paan-AG proteins have not been determined. However, comparison of the predicted amino acid sequences of these two molecules with other class I molecules showed that they are likely to have structures typical of class I molecules [14]. Both genes contain conserved amino acids that are essential in formation of the tertiary structure of class I MHC molecules, interaction with β2-microglobulin as well as those that are important in peptide

Table 1. Comparison of the human and baboon class Ib molecules[a]

Characteristic	HLA-G	Paan-AG
RNA alternatively spliced?	Yes (7 transcripts)	Yes (7 transcripts)
Tissue distribution of RNA	High in placenta, low in other tissues	High in placenta
Pre-mature stop codon in exon 6?	Yes	Yes
Transcripts encoding soluble isoforms?	Present (sHLA-G1, -G2 and -G7)	Present (sPaan-AG1, -AGx and AGxi)
Polymorphism	Limited (5 alleles)	Limited (2 alleles)
Expression of protein in placenta	Extravillous cytotrophoblast (membrane-bound) and syncytiotrophoblast (soluble isoform)	Syncytiotrophoblast and extravillous cytotrophoblast (both membrane-bound and soluble isoforms)

[a] Table modified from [8].

presentation, T-cell receptor and CD8 co-receptor binding [11, 13, 14].

Analysis of Paan-AG messages shows that they have unique characteristics that are typical of HLA-G (fig. 1, table 1). Paan-AG mRNA is alternatively spliced to form 7 transcripts and contains a premature stop codon at the beginning of exon 6, similar to HLA-G [11]. The premature stop codon results in generation of an MHC class I heavy chain with a truncated cytoplasmic domain. One of the alternatively spliced transcripts retains intron 4 and would be predicted to encode a soluble Paan-AG protein lacking the transmembrane and cytoplasmic domains. This feature (of generating soluble protein isoforms by retention of intron 4) is unique to *HLA-G* and the *AG* locus; it has not been found in other MHC class I molecules. Both genes are also relatively monomorphic, with only 5 alleles described in *HLA-G* and two in *Paan-AG* [8].

Expression Patterns of HLA-G and Paan-AG

Both HLA-G and Paan-AG exhibit a rather unique and restricted pattern of expression. HLA-G transcripts have been detected in first trimester and term villous and extravillous trophoblast cells, and in a number of non-placental tissues [8, 15]. Studies in our laboratory showed a similar pattern of expression in the baboon; Paan-AG mRNA is detectable in the placenta and in a variety of non-placental tissues [11]. All the seven alternatively spliced transcripts of Paan-AG were detected in the baboon placenta and decidua, but only two of these (the full-length transcript and one with a truncated exon 3) were detected in non-placental tissues [11]. Further analysis of the Paan-AG sequences obtained from non-placental tissues indicates that they are probably inactivated as a result of deletions in exon 3 that generate multiple stop codons and frame-shifts in the exons downstream of this deletion [11]. Thus, these transcripts may not be translated in non-placental tissues. Immunohistochemical studies using Paan-AG specific antibodies have confirmed that neither membrane-anchored nor soluble protein is expressed in non-placental tissues (Langat, DK, Hunt JS, unpublished data).

Membrane-anchored HLA-G protein is expressed constitutively in vivo by extravillous cytotrophoblasts and the thymus [6]. In addition, soluble HLA-G protein has been detected in amniotic fluid, syncytiotrophoblast and circulating blood in pregnant women [8]. HLA-G has also been detected on the surfaces of human oocytes and pre-implantation embryos [16], and it has been shown that soluble HLA-G is secreted by pre-implantation embryos [17]. In the baboon, membrane-anchored and soluble Paan-AG proteins have been detected in extravillous cytotrophoblast cells in the basal plate, similar to HLA-G [11]. However, both soluble and membrane-anchored Paan-AG proteins were also expressed in the syncytiotrophoblast in both first trimester and term placenta, in contrast to HLA-G expression, which is believed to be confined to the extravillous cytotrophoblasts. In recent experiments, we have detected alternatively spliced transcripts of Paan-AG in baboon oocytes, zygotes and blastocysts (Langat DK, Fazleabas AT, Hunt JS, manuscript in preparation). We are currently testing these samples for protein expression.

Potential Functions of HLA-G and Paan-AG

The function of HLA-G is unclear, as discussed in a number of recent reviews [6, 18, 19]. HLA-G is an oligomorphic multi-isoform molecule that can bind and present peptides similar to class Ia molecules. Studies have suggested that HLA-G may have multiple functions, including protection of the placenta from natural killer cells and activated cytotoxic T lymphocytes, induction of Th2 cytokine secretion by decidual cells [18]. In addition, soluble isoforms may act as specific immunosuppressors during pregnancy [18]. No functional studies have as yet been performed using Paan-AG, either in vitro or in vivo. We are currently in the process of generating specific monoclonal antibodies to the different Paan-AG isoforms for use in in vivo studies so as to elucidate the potential functions of Paan-AG in pregnancy.

Summary

Studies in our laboratory have shown that *Paan-AG*, a potential functional homologue of *HLA-G*, is transcribed in the baboon placenta and that the primary Paan-AG mRNA transcript is alternatively spliced in a manner similar to HLA-G mRNA. We have also shown that the transcripts are translated only in a limited number of tissues, notably the placenta, and that it is inactivated in non-placental tissues. These results closely mirror those of HLA-G in human, and strongly suggest that the *AG* locus may be a functional homologue of *HLA-G*. The current consensus is that *HLA-G* may play a critical role in pregnancy, and our results suggest the baboon would serve as an ideal animal model in experiments to evaluate this hypothesis.

Acknowledgements

This work was supported by a CONRAD Twinning grant with the Institute of Primate Research, Nairobi, Kenya, the CONRAD program (CICCR) and the University of Kansas Medical Center Research Institute, Kansas City, USA (MFG-99-44/45 to JSH and JMM), and a PO1 NIH grant (HD39878) to JSH.

References
1 Hunt JS: Immunobiology of the maternal-fetal interface. Prenat Neonat Med 1998;3:72–75.
2 Hunt JS, Johnson PM: Immunology of Reproduction; in Kuobil E, Neill JD: Encyclopedia of Reproduction. Elsevier, 1999, vol 2, pp 798–806.
3 Bainbridge DRJ: Evolution of mammalian pregnancy in the presence of the maternal immune system. Rev Reprod 2000;5:67–74.
4 Gaunt G, Ramin K: Immunological tolerance of the human fetus. Am J Perinatol 2001;18:299–312.
5 Hunt JS: Major histocompatibility complex antigens in reproduction; in Glasser S, Aplin J, Giudice L, Tabibzadeh S (eds): The Endometrium. London, Harwood Academic Publishers, 2001, pp 392–402.
6 Le Bouteiller P, Blaschitz A: The functionality of HLA-G is emerging. Immunol Rev 1999;167:233–244.
7 Bainbridge DRJ, Ellis S, Le Bouteiller P, Sargent I: HLA-G remains a mystery. Trends Immunol 2001;22:548–552.
8 Langat DK, Hunt JS: Do non-human primates provide appropriate animal models for studies on the function of HLA-G? Biol Reprod 2002;67:1367–1374.
9 Bontrop RE: Non-human primates: Essential partners in biomedical research. Immunol Rev 2001;183:5–9.
10 Castro MJ, Morales P, Fernandez-Soria V, Suarez B, Recio MJ, Alvarez M, Martin-Villa M, Arnaiz-Villena A: Allelic diversity at the primate MHC-G locus: exon 3 bears stop codons in all Cerpithecinae sequences. Immunogenetics 1996;43:327–336.
11 Langat DK, Morales PJ, Fazleabas AT, Mwenda JM, Hunt JS: Baboon placentas express soluble and membrane-bound Paan-AG proteins encoded by alternatively spliced transcripts of the class Ib major histocompatibility complex gene, *Paan-AG.* Immunogenetics 2002;54:164–173.
12 Boyson JE, Iwanaga KK, Golos TG, Watkins DI: Identification of a novel MHC class I gene, *Mamu-AG,* expressed in the placenta of a primate with an inactivated G locus. J Immunol 1997;159:3311–3321.
13 Geraghty DE, Koller BH, Orr HT: A human major histocompatibility complex class I gene that encodes a protein with a shortened cytoplasmic segment. Proc Nat Acad Sci USA 1987;84:9145–9149.
14 Bjorkman PJ, Saper MA, Samraoui B, Bennet WS, Strominger JL, Wiley DC: Structure of the human class I histocompatibility complex antigen, HLA-A2. Nature 1987;329:506–512.
15 Onno M, Guillaudeux T, Amiot L, Renard I, Drenou B, Hirel B, Girr M, Semana G, Le Bouteiller P, Fauchet R: The HLA-G gene is expressed at a low mRNA level in different human cells and tissues. Hum Immunol 1994;41:79–86.
16 Jurisicova A, Casper RF, MacLusky NJ, Mills GB, Librach CL: HLA-G expression during preimplantation human embryo development. Proc Nat Acad Sci USA 1996;93:161–165.
17 Fuzzi B, Rizzo R, Criscuoli L, Noci I, Melchiorri L, Scarselli B, Bencini E, Menicucci A, Baricordi OR: HLA-G expression in early embryos is a fundamental prerequisite for the obtainment of pregnancy. Eur J Immunol 2002;32:311–315.
18 Le Bouteiller P, Solier C, Proll J, Aquerre-Girr M, Fournel S, Lenfant F: Placental HLA-G protein expression in vivo: where and what for? Hum Reprod Update 1999;5:223–233.
19 Carosella ED, Moreau P, Aractingi S, Rouas-Freiss N: HLA-G; a shield against inflammatory aggression. Trends Immunol 2001;22:553–555.

16

The Prevalence of Anti-Phospholipid Antibodies in a Selected Population of Kenyan Women and Development of a Non-Human Primate Model

J.M. Mwenda[a], J.M. Machoki[b], E. Omollo[a], M. Galo[a], D.K. Langat[a, c]

[a] Reproductive Biology Division, Institute of Primate Research, Nairobi, and [b] Department of Obstetrics and Gynecology, University of Nairobi, Nairobi, Kenya; [c] Department of Anatomy and Cell Biology, University of Kansas Medical Center, Kansas City, KS, USA

Abstract
The mechanisms by which anti-phospholipid antibodies (aPLs) may induce pregnancy losses, intrauterine growth retardation and pregnancy-induced hypertension are not clearly understood. Moreover, there is a controversy regarding the possible direct effects of these antibodies on the physiology of the placenta since the target antigens of these antibodies are intracellular antigens and are potentially inaccessible to the antibody. Also, controversy exists regarding the usefulness of the treatment regimens currently available. In this study, we present preliminary data on the prevalence of aPLs in a selected population (n = 80) of Kenyan women visiting Kenyatta National Hospital, Nairobi, Kenya for obstetrical complications including recurrent pregnancy losses. Our results showed approximately 13.8% of the patients were positive for anti-cardiolipin antibodies whereas 33.8% were positive for aPS. Additionally, we screened 72 non-human primates for presence of aPLs and our results showed that the olive baboon *(Papio anubis)* had the highest prevalence rate (52.2%, n = 23). Overall, our results suggest that the olive baboon may be a suitable animal model for studying the mechanism of action of the anti-phospholipid antibody and pregnancy complications associated with aPLs.

Introduction
The anti-phospholipid antibody syndrome associated with the production of autoantibodies against negatively charged phospholipids (aPLs), clinically associated with thrombocytopenia, thrombosis or pregnancy losses [1]. The most common complication, however, is the disturbance of pregnancy [2]. Patients with aPLs have about an 80% pregnancy loss rate, half of which are lost in the first trimester. Even the few cases of successful pregnancy are at risk for intrauterine growth retardation, placental abruption, and early and severe pregnancy-induced hypertension [2, 3]. Therefore, the vast majority of women with high titres of aPLs have complications in pregnancy that may result in recurrent miscarriages. In our study, we report on the prevalence of aPL in a small population (n = 80) of Kenyan women referred to Kenyatta National hospital, Nairobi, Kenya for obstetrical complications including recurrent pregnancies losses. In addition, we also present data from our studies at the Institute of Primate Research, Nairobi, Kenya, suggesting that the olive baboon *(Papio anubis)* may be a good model to study the mechanism of action of aPLs on the placenta and discusses the potential of using this model to evaluate the safety and efficacy of various drugs for treatment of pregnancy complications associated with aPLs.

The Anti-Phospholipid Syndrome
The most commonly reported aPLs are the lupus anti-coagulant (LA), antibodies against cardiolipin (aCL) and phosphatidylserine

(aPS) [4–6]. The LA is identified functionally by prolongation of in vitro PL-dependent coagulation tests. ACL and aPS are identified by ELISAs or binding to phospholipid-coated glass microspheres [7, 8].

Potential Mechanism of aPL That May Lead to Pregnancy Complications

Previous reports suggest that aPLs can directly affect the placental trophoblasts, leading to inadequate trophoblastic differentiation, incomplete syncytium formation, suppression of decidual invasion, or loss of endovascular trophoblast control of maternal coagulation [3, 5, 6]. This conclusion arose from the observation that aPS and aCL antibodies are the prevalent antibodies generally associated with the aPL antibody syndrome. Phosphatidylserine is a predominant phospholipid on the inner side of plasma membranes and under certain normal physiologic conditions; it is externalized and therefore exposed to the circulating aPS [9]. Phosphatidylserine is also externalized during the intercellular fusion and differentiation process in normal placental development [1]. It has been shown that aPLs can bind to externalized phosphatidylserine and in vitro cause incomplete differentiation, block fusion, and inhibit invasion. Thus, aPS antibodies can access the antigen in such situations [5, 6, 10].

However, it is not possible to explain the commonly observed association between recurrent pregnancy loss, aCL and LA. Although the criteria for diagnosis of the aPL syndrome, LA and aCL assays have been standardized, there are discrepancies between the laboratory and the clinical features of patients considered to have aPL syndrome [8]. It is also difficult to explain how aCL, usually leads to pathophysiology of the placenta. This is because cardiolipin has a restricted distribution to the inner mitochondrial membrane while phosphatidylcholine, phosphatidylethanolamine and phosphatidylserine are the principal phospholipid components of the plasma membrane. Since antibodies do not freely cross the plasma membranes of living cells, aCL therefore cannot access its corresponding antigen.

Prevalence of aPL in a Selected Population of Kenyan Women

Most of the available studies on the prevalence of aPLs in women were mainly carried out in the United States and Europe. There is very little data on the prevalence of aPLs in African women. In this regard, we initiated pilot studies to document the prevalence of aPLs in African women. The samples were assayed by ELISA as previously described [11] using kits kindly donated by REAADS Medical Products, Westminster Inc, Colorado, USA and used for semiquantitative determination of IgG and IgM anti-cardiolipin antibodies. These samples were collected from women (n = 80) referred to Kenyatta National Hospital in Nairobi, Kenya for reasons related to obstetrical complications including recurrent pregnancy losses. This hospital is the biggest referral hospital in East and Central Africa, thus data obtained from this hospital would be representative of the population in this region.

Results

Prevalence of aPLs in a Selected Kenyan Population

Our results showed approximately 13.8% of the patients were positive for aCL and negative for aPS, 33.8% were positive for aPS and aCL negative and 28.8% were positive for both aCL and aPS whereas 23.8% patients were negative for both aPS and aCL (table 1, 2).

Table 1. Prevalence of aPLs in a selected population of Kenyan women

Patients (n = 80)	aPS positive	aPS negative
aCL positive	28.75% (23)	13.75% (11)
aCL negative	33.75% (27)	23.75% (19)

Table 2. Prevalence of aPL antibodies (by isotype) in Kenyan women

Patients (n = 80)	APS		aCL	
	IgG	IgM	IgG	IgM
Positive	18 (22.5%)	34 (42.5%)	23 (28.8%)	28 (35%)

The Prevalence of aPLs Female in Baboons and Development of a Non-Human Primate Model

There are a lot of unanswered questions regarding the role of aPLs in pregnancy losses, including the mechanism of action. Some investigators have questioned the diagnostic value of using aPL assays as an indicator of underlying autoimmune disease or potential pathological complications that may be associated with aPLs in an on-going pregnancy. This is mainly because of observations that some women with aPLs have successful pregnancies. Due to ethical concerns that place limitations on the use of humans for medical research, animal models especially non-human primates (NHP) offer suitable alternatives. NHPs are phylogenetically close to humans and would be the most ideal models for these studies. In particular, female baboons have similarities with women especially in relation to reproductive physiology, anatomy, implantation and pregnancy events.

Thus, our studies have focused on the development of the baboon *(Papio anubis)* as a NHP model for aPL. In a preliminary study conducted in our laboratory, a total of 72 non-human primates of seven different species were assessed for presence of anti-phosphatidylserine (aPS) IgG antibodies using enzyme-linked immunosorbent assay (ELISA). These animals had a history of recurrent abortion or stillbirths previously. We found that almost half of the animals tested (33 out of 72, 45.83%) were positive for aPS antibodies [11]. The olive baboon had the highest prevalence rate (52.2%, n = 23) followed by the African green monkey *(Cercopithecus aethiops)* (47.18%, n = 23).

Immunohistochemical Localization of aPL-Cross-Reactive Antigens in Baboon Placenta

We used monoclonal antibodies produced against CL and PS-dependent antigens and determined immunohistochemical reactivity with baboon placental tissues of varying gestational ages. The antibody BA3 had previously been shown to react with both PS- and CL-dependent antigens in routine clinical ELISAs whereas 3SB reacts with PS-dependent antigens [12]. D11 reacts with CL-dependent, but not PS-dependent antigens.

In 12-week and 24-week baboon placental tissue sections, BA3 showed intense labeling of the cytotrophoblast cell layer. In the 12-week tissue sections, BA3 strongly labeled the area of the developing cytotrophoblasts, but did not label the area of developing syncytiotrophoblasts (Mwenda JM, Langat DK, unpublished data). BA3 strongly labeled the individual cytotrophoblast cells and weakly stained areas of the syncytiotrophoblasts in the 24-week tissue sections. The baboon samples labeled significantly stronger with BA3 than human samples. BA3 labeled both human and baboon tissues with the same pattern. BA3 strongly labeled the cytotrophoblasts in both tissue samples, but the baboon syncytiotrophoblast layer was faintly stained in the 24-week baboon tissue.

3SB showed a strong labeling of the syncytiotrophoblast cell layer in both 12-week and 24-week baboon placental tissue sections. In the 12-week tissue sections, 3SB strongly labeled the layer of developing syncytiotrophoblasts. In the 24-week samples, 3SB strongly labeled the syncytiotrophoblast layer in some areas and labeled weakly in others. Baboon and human placental samples showed a very similar reactivity to 3SB as the syncytiotrophoblast layer was strongly labeled in both. D11 showed no labeling in both 12-week and 24-week baboon placental tissue sections, similar to what was observed in human placental tissue sections. No area of the villi reacted to D11.

Summary

Analysis of the immunological data results obtained from the baboon samples showed a pattern of expression of anti-phospholipid cross-reactive antigens in baboon placental tissues similar to that of human placenta. Thus, we have showed that a significant percentage of non-human primates with a history of recurrent pregnancy losses have circulating aPS as shown by ELISA assays [11]. The ELISA results were consistent with immunohistochemical reactivity in baboon and human placental tissues. Taking into account the similarity of the reproductive physiology (and other advantages of the baboon a model for reproductive health studies) of the baboon and humans, it has become apparent that the baboon could be an ideal model for the study of the anti-phospholipid syndrome and its management. The presence of naturally occurring aPLs in non-human primates suggest that the baboon model could be valuable in elucidating the mechanism of action aPLs and evaluation of safety and efficacy of new therapeutic agents for treatment of early pregnancy losses caused by aPLs. Also, we have shown that the Kenyan women have naturally occurring antibodies to aPLs (aCL and aPS antibodies) and these antibodies may have caused the recurrent pregnancy losses in this selected population of Kenyan women.

Acknowledgment

We thank Prof. Neal Rote PhD, Director of the Division of Research, Department of Obstetrics and Gynecology, MetroHealth Medical Center, Case Western Reserve University School of Medicine, Cleveland, OH and Dr. Mike Kent, Neuroscience Institute, Dayton, USA for kindly donating the ELISA kits and anti-phospholipid antibodies and protocols used in this study. Also, we would like to thank Prof C. Sekadde-Kigondu, Department of Clinical Chemistry, University of Nairobi, Kenya, for providing some of the samples used in this study.

References

1 Rote NS: Anti-phospholipid antibodies and recurrent pregnancy loss. Am J Reprod Immunol 1996;35:394–401.
2 Sammaritano LR: Update on the management of the pregnant patient with antiphospholipid antibody. Curr Rheumatol Rep 2001;3:213–221.
3 Rote NS, Walter A, Lyden TW: Anti-phospholipid antibodies: Lobsters or red herrings? Am J Reprod Immunol 1992;28:31–37.
4 Alving BM: Diagnosis and management of patients with the antiphospholipid syndrome. J Thrombosis Thrombolysis 2001;12:89–93.
5 Triplett DA: Antiphospholipid antibodies. Arch Pathol Lab Med 2002;126:1424–1429.
6 Durrani OM, Gordon C, Murray PI: Primary anti-phospholipid antibody syndrome (APS): current concepts. Surv Ophthalmol 2002;47:215–38.
7 Loizou S, McCrea JD, Rudge AC, Reynolds R, Boyle CC, Harris EN: Measurement of anti-cardiolipin antibodies by an enzyme-linked immunosorbent assay (ELISA): standardization and quantitation of results. Clin Exp Immunol 1985;62:738–745.
8 Branch DW: Anti-phospholipid antibodies and reproductive outcome: the current state of affairs. J Reprod Immunol 1998;38:75–87.
9 Rote NS, Vogt E, De Vere G, Obringer AR, Ng AK: The role of placental trophoblast in the pathophysiology of the anti-phospholipid antibody syndrome. Am J Reprod Immunol 1998;39:125–136.
10 Kutteh WH, Rote NS, Silver R: Antiphospholipid antibodies and reproduction: the antiphospholipid antibody syndrome. Am J Reprod Immunol 1999;41:133–152.
11 Langat DK, Sichangi MW, Rote NS, Kent M, Omollo EO, Chai D, Njenga MN, Nyaundi JK, Mwenda JM: The prevalence of anti-phospholipid antibodies and fetal loss in a non-human primate colony in Kenya. J Obstet Gynecol East Central Africa 1998;14:55–61.
12 Rote NS, Ng AK, Dostal-Johnson DA, Nicholson SL, Siekman R: Immunologic detection of phosphatidylserine externalization during thrombin-induced platelet activation. Clin Immunol Immunopathol 1993;66:193–200.

17
Pre-Eclampsia and Vascular Activation in Women and Non-Human Primates

R.N. Taylor[a], C.J.M. de Groot[b]

[a] Center for Reproductive Sciences, Department of Obstetrics, Gynecology and Reproductive Sciences, University of California, San Francisco, San Francisco, Calif., USA;
[b] Department of Obstetrics, Gynecology and Reproductive Medicine, Erasmus University Medical Centre, Rotterdam, The Netherlands

Pre-eclampsia (PE) is a pregnancy-specific syndrome defined clinically by hypertension, edema and proteinuria. It affects 4–7% of pregnant women worldwide and in the United States, PE is the third leading cause of maternal mortality [1]. The etiopathogenesis of PE is understood poorly. Consequently its therapy, including antihypertensive medication, seizure prophylaxis and expedient delivery, is based on the effects rather than the causes of the syndrome. These empirical interventions, particularly emergent operative and often preterm delivery, result in unacceptably high rates of neonatal morbidity and mortality.

So many hypotheses have been advanced to explain the pathogenesis of this PE that it has been referred to as the 'disease of theories'. Over the past decade attention has been focused on the triad of uteroplacental ischemia, impaired trophoblast invasion and generalized maternal endothelial cell activation as mechanistic factors in PE pathogenesis [2]. In a recent review Brosens et al. [3] reassess their classical histochemical studies of the placental beds of preeclamptic women and suggest that faulty preparation of the endometrium may underlie this and other placental defects. Developing human trophoblasts progress through an 'intermediate' phase of differentiation [4] and subsequently are committed to fusion and terminally differentiate to secretory syncytiotrophoblast or are recruited to an invasive phalanx to establish an anchoring villus [5]. The failure of cytotrophoblasts to matriculate along the invasive differentiation pathway leads to trophoblast proliferation and the formation of syncytial knots that are hallmarks of placental bed pathology in PE [6].

In recent years it has become widely accepted that endothelial cell dysfunction is a key component of the PE syndrome [7]. This brief review will focus on the evidence supporting the involvement of endothelial cell activation in its pathophysiology. Spargo and colleagues were the first to define the pathognomonic histology of reversible endothelial cell damage, 'glomeruloendotheliosis,' in the kidneys of women with PE [8]. Very similar pathology has been observed in PE-like syndromes occurring spontaneously in the pregnant baboon [9] and in an experimental rhesus model of PE described below [10, 11].

Endothelial cells are positioned at the interface between circulating blood and vascular smooth muscle or the extravascular space, where they occupy a surface area of more than 1,000 m^2. These cells secrete a variety of signaling molecules directly into the circulation. In turn, endothelial cells are themselves targets for cellular and soluble plasma constituents (e.g., platelets, leukocytes, fragments of placental membranes, cytokines, antibodies and other circulating peptides). Endothelial cells modulate the regulation of vascular tone, coagulation, permeability, and the targeting of immune cells. Vascular endothelial cell 'activation' is a term used to define an altered state of endothelial cell differentiation, typically induced as a result of cytokine stimulation [12]. It often represents a response to sublethal injury of these cells and is thought to play a major role in the pathophysiology of atherosclerosis and PE. Endothelial cell dysfunction in PE may result from a variety of factors, including physical shear forces, hypoxia, lipid peroxides in addition to the circulating constituents described above.

In PE, endothelial activation is manifested functionally by increased sensitivity to pressor agents [13] and resistance to vascular flow [14]. Biochemical evidence of maternal endothelial cell activation in PE includes reduced vasodilator levels (e.g., prostacyclin [15]) and increased vasoconstrictor concentrations (e.g., endothelin-1), as well as elevated concentrations of circulating extracellular matrix proteins and adhesion molecules, including ICAM-1, VCAM-1, and P-selectin [16]. Recent studies indicate that the predictive values of some of these tests may approach levels that would be clinically useful [17, 18]. When these changes are initiated in response to acute mechanical or biochemical endothelial cell damage they promote efficient wound healing. However, when activated by a chronic pathological process, such as PE, the same responses can create a vicious cycle of vasospasm, microthrombosis, disruption of vascular integrity and serious physiologic disturbances until the inciting factor is eliminated.

In response to injury or activation, endothelial cells express two high molecular weight glycoproteins with procoagulant and adhesion properties, fibronectin and von Willebrand factor (vWF). Both proteins are predominantly localized to the abluminal extracellular matrix of human endothelium [19], but as discussed above these can be actively secreted from endothelial cells under conditions of cellular activation. Degraded fibronectin is chemoattractant for neutrophils, and proteolytic activity from these cells furthers fibronectin breakdown [20].

Elevated concentrations of fibronectin in the plasma of women with PE have been recognized for over 15 years [21]. The endothelial isoform, cellular fibronectin (cFN), expresses two extra domains (ED-A and ED-B) generated by differential mRNA splicing. While it is a major component of the endothelial extracellular matrix, cFN is normally only a minor component of circulating fibronectin and thus is an accurate marker of endothelial cell injury [22, 23]. By contrast, women with transient hypertension, a hypertensive condition of pregnancy without proteinuria and without the maternal or neonatal morbidity associated with preeclampsia, had normal cFN levels.

The data described thus far indicate that PE is associated with elevated circulating concentrations of endothelial cell biomarkers. To test the hypothesis that placental ischemia might cause endothelial cell activation, pregnant rhesus macaques at 119 ± 1 days gestation were subjected to partial aortic ligature distal to the renal arteries or sham surgery. The ligature resulted in a mean 25 mm Hg decrease in uterine perfusion pressure [11]. At 152 ± 2 days all the animals were delivered by caesarean section. Of 8 experimental monkeys, five developed systemic hypertension and five developed proteinuria. None of the sham operated controls had either manifestation. At the time of delivery, uterine, renal and rectus muscle biopsies also were performed and analyzed by semiquantitative immunohistochemistry. To assess vascular endothelial activation in these tissues, antibodies against vWF were utilized. The results demonstrated that vWF expression in the endothelia of these tissues was significantly increased in the operated monkeys (two-factor ANOVA, $p = 0.04$). The findings support the postulate that impaired placental perfusion can cause systemic maternal vascular endothelial cell activation in primates.

Summary

The PE syndrome is a complex and multi-faceted disorder of primate pregnancy. Failure of cytotrophoblast invasion into the maternal decidua appears to be a primary defect. However, while placental pathology appears to be necessary, it is not sufficient for clinically manifested preeclampsia. A secondary phase of the syndrome, in which systemic maternal endothelial cell dysfunction forms a final common pathway, must be acquired. Recent evidence suggests that the pathogenesis of the endothelial phase of preeclampsia involves factors derived from the ischemic placenta. The precise nature of these factors, cytokines, membrane fragments or oxidized lipids, remains an area of intensive investigation. Maternal constitutional factors, particularly adipose mass and general vascular health also are involved. The interaction between initiating events (i.e., trophoblast invasion, relative placental ischemia) and maternal constitutional factors (i.e., genetic disposition, total body fat, antioxidant reserves) give rise to the syndrome of PE. It is hoped that primate models of this fascinating disease will be further developed to aid discovery of effective treatment and prevention of PE.

Acknowledgement

The studies were supported in part by the U.S. National Institutes of Health through grants P01-HD24180, P01-HD30367, and R01-HL64145.

References

1 Berg CJ, Chang J, Callaghan WM, et al: Pregnancy-related mortality in the United States, 1991–1997. Obstet Gynecol 2003;101:289–296.
2 de Groot CJ, Taylor RN: Preeclampsia: an update. Eur J Obstet Gynecol Reprod Biol 1996;69:59–60.

3 Brosens JJ, Pijnenborg R, Brosens IA: The myometrial junctional zone spiral arteries in normal and abnormal pregnancies: a review of the literature. Am J Obstet Gynecol 2002;187:1416–1423.
4 Kurman RJ: The morphology, biology, and pathology of intermediate trophoblast: a look back to the present. Hum Pathol 1991, 22:847–855.
5 Norwitz ER, Schust DJ, Fisher SJ: Implantation and the survival of early pregnancy. N Engl J Med 2001 Nov 8;345:1400–1408.
6 Redline RW, Patterson P: Pre-eclampsia is associated with an excess of proliferative immature intermediate trophoblast. Hum Pathol 1995;26: 594–600.
7 Roberts JM, Taylor RN, Musci TJ, et al: Preeclampsia: an endothelial cell disorder. Am J Obstet Gynecol 1989;161:1200–1204.
8 Spargo BH, Lichtig C, Luger AM, et al: The renal lesion in preeclampsia. Perspect Nephrol Hypertens 1976;5:129–137.
9 Hennesey A, Gillin AG, Painter DM, et al: Evidence for preeclampsia in a baboon pregnancy with twins. Hypertens Pregnancy 1997;16:223–228.
10 Combs CA, Katz MA, Kitzmiller JL, et al: Experimental preeclampsia produced by chronic constriction of the lower aorta: validation with longitudinal blood pressure measurements in conscious rhesus monkeys. Am J Obstet Gynecol 1993;169:215–223.
11 de Groot CJM, Merrill DC, Taylor RN, et al: Increased von Willbrand factor expression in an experimental model of preeclampsia produced by reduction of uteroplacental prefusion pressure in conscious rhesus monkeys. Hypertens Pregnancy 1997;16:177–185.
12 Pober JS, Cotran RS: Cytokines and endothelial cell biology. Physiol Rev 1990;70:427–451.
13 Gant NF, Daley GL, Chand S, et al: A study of angiotensin II pressor response throughout primigravid pregnancy. J Clin Invest 1973;52:2682–2689.
14 Takata M, Nakatsuka M, Kudo T: Differential blood flow in uterine, ophthalmic, and brachial arteries of preeclamptic women. Obstet Gynecol 2002;100:931–939.
15 Mills JL, DerSimonian R, Raymond E, et al: Prostacyclin and thromboxane changes predating clinical onset of preeclampsia: a multicenter prospective study. JAMA 1999;282:356–362.
16 Taylor RN, de Groot CJ, Cho YK, et al: Circulating factors as markers and mediators of endothelial cell dysfunction in preeclampsia. Semin Reprod Endocrinol 1998;16:17–31.
17 Chavarria ME, Lara-Gonzalez L, Gonzalez-Gleason A, et al: Maternal plasma cellular fibronectin concentrations in normal and preeclamptic pregnancies: a longitudinal study for early prediction of preeclampsia. Am J Obstet Gynecol 2002;187:595–601.
18 Krauss T, Emons G, Kuhn W, et al: Predictive value of routine circulating soluble endothelial cell adhesion molecule measurements during pregnancy. Clin Chem 2002;48:1418–1425.
19 Aznar-Salatti J, Bastida E, Buchanan MR, et al: Differential localization of von Willebrand factor, fibronectin and 13-HODE in human endothelial cell cultures. Histochemistry 1990;93:507–511.
20 Forsyth KD, Levinsky RJ: Fibronectin degradation; an in-vitro model of neutrophil mediated endothelial cell damage. J Pathol 1990;161:313–319.
21 Lazarchick J, Stubbs TM, Romein L, et al: Predictive value of fibronectin levels in normotensive gravid women destined to become preeclamptic. Am J Obstet Gynecol 1986;154:1050–1052.
22 Lockwood CJ, Peters JH: Increased plasma levels of ED1+ cellular fibronectin precede the clinical signs of preeclampsia. Am J Obstet Gynecol 1990;162:358–362.
23 Taylor RN, Crombleholme WR, Friedman SA, et al: High plasma cellular fibronectin levels correlate with biochemical and clinical features of preeclampsia but cannot be attributed to hypertensionalone. Am J Obstet Gynecol 1991;165:895–901.

Endometriosis: Baboon Model and in vitro Models

18

The Prevalence of Endometriosis among African-American and African-Indigenous Women

M.C. Kyama[a, b], *T.M. D'Hooghe*[b], *S. Debrock*[b], *J. Machoki*[c], *D.C. Chai*[a], *J.M. Mwenda*[a]

[a] Institute of Primate Research, Karen and Nairobi, Kenya; [b] Department of Obstetrics and Gynaecology, University Hospital Gasthuisberg, Leuven, Belgium; [c] Department of Obstetrics and Gynaecology, University of Nairobi, Kenya

Abstract

Endometriosis is a gynaecological disorder, characterized by the growth of endometrial tissue outside the uterine cavity. It is the most common cause of pelvic pain and occurs in 20–25% of women with infertility. Although Sampson first described endometriosis in 1927, studies on the prevalence of endometriosis among African women are still lacking. The current thinking is that endometriosis rarely affects women from African origin. However, in African-American women in the USA, endometriosis is one of the commonest indications for major gynaecological surgery and hysterectomy, and is associated with long hospital stay and high hospital charges. There is also some evidence that endometriosis is more commonly found in African-American patients from private practice than in African-American patients treated in public hospitals. The prevalence of endometriosis in African-indigenous women with infertility seems low, possibly due to a different life style (early pregnancy, increased risk for PID and blocked Fallopian tubes) and due to lack of laparascopic facilities and specific training of African gynecologists to diagnose this disease. Specific manifestations like cervical endometriosis and ascites caused by endometriosis appear to be more frequently observed in African-indigenous or African-American women than in women with other ethnic backgrounds.

Introduction

Endometriosis, a disease that affects at least 10% of menstruating women, is characterized by growth of endometrial tissue outside the uterine cavity. It is the most common cause of pelvic pain and occurs in 13–33% of women with infertility [1]. Endometriosis has been erroneously referred to as the *career women's disease*, probably because working women had more resources to seek continuous medical care. A neglected subject of investigation is the difference in the prevalence of endometriosis among different races. Although the prevalence of endometriosis is well documented in women living in developed countries, studies on the prevalence of this disease among African women are still lacking. Some clinical investigations have shown that the prevalence of endometriosis may be lower in black women and higher in oriental women than in the Caucasian population [2, 3]. However, the designation of race in itself is problematic, and its usefulness as an indicator of genetic or biological factors in a population can be challenged, since race is frequently a marker for cultural, economic and medical factors [4]. Nevertheless, there is a general perception that endometriosis rarely occurs among black women. This is based on the hypothesis that pregnancy at early age, early marriage and high incidence of pelvic inflammatory disease tend to preclude the development of endometriosis. In this short paper, we demonstrate that endometriosis may be more common in African-indigenous women (born and living in Africa) and in African-American women (born and living in North America) than is generally accepted. It is critical to carry out more studies to elucidate the prevalence and the dynamics of this disease in women from African origin.

Hysterectomy Indications

Endometriosis was the indication for hysterectomy in 9% of black women and 20% of white women in the USA during the period 1988–1990 [4]. The estimated rate of hysterectomies for endometriosis per 10,000 women aged 15 years or older in the US civilian population is 5.5 in black women and 11.2 in white women. In comparison, uterine leiomyoma accounted for 61% of hysterectomies for black women and 29% for white women [4]. Furthermore, the prevalence and cost of endometriosis-related hospitalisations has been determined in a retrospective analysis based on nationwide clinical practice data in the USA [5]. Most endometriosis admissions occurred in women aged 35–49 and the most common procedure was a total abdominal hysterectomy (55–60%). Older and African-American patients had the longest length of hospital stay and the highest total charges. The estimated total hospitalisation costs, as represented by the hospital charges, for women with endometriosis as the primary diagnosis in the USA were USD 540 million for 1991 and USD 579 million for 1992, suggesting that endometriosis-related hospitalisation is a major burden on healthcare systems [5]. Similarly, a review of all private patients admitted for major gynaecological surgery to a private US clinic [6] indicated that the prevalence of endometriosis was similar in African-American women (6.9%) and in white women (7.7%). Collectively, these data (table 1) clearly show that endometriosis is a common indication for major gynaecological surgery in African-American women. In Nigeria, endometriosis has been found in only 0.2% [7] or 0.4% [8] of women undergoing gynaecological operations, excluding laparoscopy (table 1).

Pain and Infertility

In infertile Caucasian women the prevalence of endometriosis in infertile Caucasian women has been reported to vary between 13% in the period 1970–1987 and 33% in the period 1988–2000 [1].

In African-American women [9], the laparascopic prevalence of endometriosis in African-American patients aged 18–40 years with pain, infertility, menstrual disorders or benign gynecological conditions was 23%, suggesting that endometriosis is also a frequent condition in black women of younger age than those studied in the hysterectomy studies mentioned above. In a later report [10], the prevalence of endometriosis in 43 adolescent African-American women aged 18–19 years presenting with pain or abnormal vaginal bleeding was 65%, including mild (50%), moderate (39%) or severe (11%) endometriosis according to the Acosta classification [11]. In a more recent study [12], the prevalence of endometriosis in infertile women in Cincinatti, USA (table 1), was low but comparable in white patients (4.7%) and in black patients (2.6%).

In African-indigenous women at the Fertility Clinic at Groote Schuur Hospital in Cape Town, South Africa [13], the prevalence of endometriosis was lower in black patients (2%) than in colored patients (4%) or white patients (6%). In Nigeria, the prevalence of endometriosis in infertile black patients was apparently in the same range: 1.8% [14] and 1.4% [15].

Table 1. Prevalence of endometriosis in women from African origin according to symptoms and surgical procedures

Symptoms/operations	Ethnic groups		
	African-indigenous, %	African-American, %	Caucasian %
Infertility	1.8 [14] 1.4 [15] 2 [13]	2.6 [12]	4.7 [12] 13–33 [1]
Infertility/pain/abnormal bleeding/benign gynecology	–	65 [10] 23 [9]	15 [10]
Hysterectomy	–	9 [4]	20 [4]
Gynecological surgery excluding laparoscopy	0.2 [7] 0.4 [8]	6.9 [6]	7.7 [6]

Numbers in brackets correspond to references.

Underreporting of Endometriosis in African-Indigenous Women?

Overall, it seems that endometriosis may be more commonly found in infertile Caucasian or African-American women than in African-indigenous women. Obviously, this difference can be also related to non-racial factors protecting against endometriosis, since it is known that African-indigenous women have children at an earlier age, have more children than Caucasian women in developed countries or than African-American women, and have tubal infertility as leading cause (60%) of female infertility [13]. All these factors limit the cumulative number of menstrual cycles with retrograde menstruation that is positively correlated with the risk of endometriosis [16]. It would be interesting to study the prevalence of endometriosis in parous women of African origin with secondary infertility and control for the number of menstrual cycles since their last delivery. Furthermore, it is very likely that the true prevalence of endometriosis in African-indigenous women is underreported due to the following reasons [1]. Firstly, the prevalence may vary with the diagnostic method used: laparoscopy is generally accepted to be a better method for the diagnosis of minimal to mild endometriosis than laparotomy and this technique has only become widely available during the last 15–20 years in the western world, and is not available in many African countries. Secondly, most African gynecologists have limited training, and experience in diagnosing endometriosis by laparoscopy and in recognizing the wide variability in appearance of especially subtle endometriosis implants, cysts and adhesions is important. The concept of subtle endometrial lesions and deep endometriosis has only been reported after 1985, and is not yet fully appreciated by all gynecologists. Thirdly, most studies evaluating the prevalence of endometriosis in women of reproductive age lack systematic histological confirmation of the macroscopic findings. However, pathological confirmation of the laparoscopic impression is essential for the diagnosis of endometriosis [1].

Specific Presentations

Cervical endometriosis and endometriosis-related ascites appear to be two specific manifestations of endometriosis that are more common in African-American and/or African-indigenous women than in Caucasian women. In a series of 570 African-American patients undergoing colposcopic evaluation for abdominal cervical smears, vaginal discharge and abnormal vaginal bleeding [17], the prevalence of cervical endometriosis was 2.4% (16/570). Almost 56% of these patients with cervical endometriosis had previously documented cervical procedures, including cone biopsy, cervical curretage, laser surgery, or electrocautery [17]. None of them had evidence of internal pelvic or extrapelvic endometriosis on clinical examination, but a laparoscopy was not performed to exclude the presence of pelvic endometriosis. Cervical endometriosis needs to be considered as a possible cause of intermenstrual bleeding, perimenstrual spotting or contact bleeding [17].

Ascites due to endometriosis is a rare manifestation of endometriosis [18]. In a recent review, 20 cases were reported and 82% were nulliparous young black women [18]. The main presenting symptoms of endometriosis-associated acites include abdominal distention and pain (chronic or dysmenorrhea). Whereas pelvic endometriosis was found in all patients, some patients also had omental or pleural endometriosis [18]. It has been hypothesized that rupture of chocolate cysts in the abdominal cavity with release of endometrial cells and blood causes peritoneal irritation leading to ascites [18].

Other extrapelvic manifestations of endometriosis including umbilical endometriosis [20] and pulmonary endometriosis [21] have also been reported in women of African origin.

Conclusions

Endometriosis appears to be an important indication for major gynecological surgery and a significant cause of infertility in African-American women. The prevalence of endometriosis in African-indigenous women has not been studied adequately, but appears to be lower, possibly related to early and/or high parity, tubal infertility and underdiagnosis by gynecologists. The impact of a merely racial factor is probably limited, since the prevalence of endometriosis appears to be higher in African-American women than in African-indigenous women (table 1). African-indigenous women are currently experiencing a fast change in both lifestyle and social economic status. As they marry later, have their first child at an older age, and have less children, a higher amount of menstrual cycles with retrograde menstruation will most likely increase their risk to develop endometriosis [16]. Therefore, endometriosis should be considered as a potential cause of pelvic pain and subfertility in all women, including African-indigenous women. More detailed studies are required on the prevalence, laparascopic appearance and histological confirmation of endometriosis among African-American and African-indigenous women with infertility, pain or other gynaecological problems.

References

1 D'Hooghe TM, Debrock S, Hill JA, Meuleman C: Endometriosis and subfertility: is the relationship resolved? Semin Reprod Med 2003, in press.
2 Miyazawa K: Incidence of endometriosis among Japanese women. Obstet Gynecol 1976;48:407–409.
3 Arumugam K, Templeton AA: Endometriosis and Race. Aust NZ J Obstet Gynecol 1992;32:164–165.
4 Wilcox LS, Koonin LM, Pokras R, Strauss LT, Xia Z, Peterson HB: Hysterectomy in the United States, 1988–1990. Obstet Gynecol 1994;83:549–555.
5 Zhao SZ, Wong JM, Davis MB, Gersh GE, Johnson KE: The cost of inpatient endometriosis treatment: an analysis based on the the Healthcare cost and utilization project Nationwide Inpatient sample. Am J Manag Care 1998;8:1127–1134.
6 Lloyd FP: Endometriosis in the Negro woman. Am J Obstet Gynecol 1964; 89:468.
7 Chatman DL: Endometriosis and the black woman. J Reprod Med 1976; 16:303–306.
8 Ekwempu CC, Harrison KA: Endometriosis among the Hausa/Fulani population at Nigeria. Trop Geogr Med 1979;31:201–205.
9 Aimakhu VE, Osunkayo BO: Endometriosis externa in Ibadan, Nigeria. Am J Obstet Gynecol 1971;110:489–493.
10 Chatman DL, Ward AB: Endometriosis in adolescents. J Reprod Med 1982;27:156–160.
11 Acosta AA, Buttram VC Jr, Besch PK, et al: A proposed classification of pelvic endometriosis. Obstet Gynecol 1973;42:19.
12 Green JA, Robbins JC, Scheiber M, Awadalla S, Thomas MA: Racial and economic demographics of couples seeking infertility treatment. Am J Obstet Gynecol 2001;184:1080–1082.
13 Wiswedel K, Allen DA: Infertility factors at the Groote Schuur Hospital. S Afric Med J 1989;76:65–66.
14 Thacher TD, Nwana EJC, Karshima JA: Extrapelvic endometriosis in Nigeria. Int J Gynecol Obstet 1997;57:57–58.
15 Otolorin EO, Ojengbede O, Falase, AO: Laparoscopic evaluation of the tuboperitoneal factor in infertile Nigerian women. Int J Gynecol Obstet 1987;25:47–52.
16 D'Hooghe TM, Debrock S: Endometriosis, retrograde menstruation and peritoneal inflammation. Hum Reprod Update 2002;8:84–88.
17 Veiga-Ferreira MM, Leiman G, Dunbar F, Magolius KA: Cervical endometriosis: facilitated diagnosis by fine needle aspiration cytogenic testing. Am J Obstet Gynecol 1987;4:849–856.
18 Spitzer M, Benjamin F: Ascites due to endometriosis. Obstet Gynecol Surv 1995;50:628–631.
19 Berstein JS, Perlow V, Brenner JJ: Massive ascites due to endometriosis. Am J Digest Dis 1961;6:1.
20 Blumenthal NJ: Umbilical endometriosis. A case report. S Afric Med J 1981;59:198–199.
21 Margolis MT, Thoen LD, Mercer LJ, Keith LG: Hemothorax after Lupron therapy of patient with pleural endometriosis. A case report and literature review. Int J Fertil Menop Stud 1996;41:53–55.

19

Etiology of Endometriosis: Hypotheses and Facts

G.A.J. Dunselman, P.G. Groothuis

Department of Obstetrics and Gynaecology, Research Institute GROW, Maastricht University, Maastricht, The Netherlands

Introduction

Endometriosis may be defined as a disease characterised by the presence of functional endometrial glands and stroma in ectopic locations outside the uterine cavity. The ectopic endometrial tissue responds to ovarian steroids in a way similar to eutopic endometrium. Endometriosis is linked to the reproductive years, it does not occur in premenarcheal girls and is believed to regress in the postmenopausal years.

The first histological description of a lesion consistent with endometriosis was given by Von Rokitanski in 1860. The term Endometriosis was first used by Sampson in 1921. Since the classic papers of Sampson on the pathogenesis of endometriosis in the years between 1921 and 1940 an overwhelming number of scientific papers has been published on nearly every aspect of the disease.

Nevertheless, despite extensive basic and clinical research, endometriosis remains a disease without a precise definition, it has an obscure pathogenesis and etiology, an extremely variable clinical presentation, an unpredictable course, and, except for surgical or physiological menopause, no known cure. Endometriosis tends to recur regardless of the treatment given.

Pathogenesis

Two main hypotheses exist on the pathogenesis of endometriosis: (1) in situ development, metaplasia, induction, or development from Mullerian remnants and (2) transplantation of endometrium. According to the hypothesis of metaplasia, endometriosis arises as a result of metaplasia of the peritoneal serosa, secondary to inflammatory stimuli or hormonal influences. Based on studies in the rabbit it was suggested that substances liberated by endometrium could induce endometriosis like lesions in the undifferentiated mesenchyme. More recently it was shown that ovarian endometriotic lesions are able to arise as process of metaplasia from ovarian surface epithelial/mesothelial cells in the presence of endometrial stromal cells and estradiol.

The concept of endometriosis as a transplantation phenomenon involves different routes of dissemination. Iatrogenic, lymphogenic and hematogenic spread account for the rare, extraperitoneal, locations of endometriosis. A more obvious route of dissemination is by the Fallopian tubes. According to the theory of implantation by retrograde menstruation endometriosis is the consequence of reflux of endometrial fragments through the Fallopian tubes during menstruation with subsequent implantation and growth on and into the peritoneum. The reflux implantation theory is based on the assumption that retrograde menstruation takes places and that viable endometrial tissue reaches the abdominal cavity and implants. The first two assumptions have proven to be true, the third one, the actual implantation of endometrial tissue on the peritoneal surface has not been observed in women. Despite numerous biopsies of the peritoneal surface taken in women with endometriosis, the first contact of a fragment of endometrial tissue with the peritoneal lining or with the subperitoneal extracellular matrix has never been observed. The facts: (1) We know that in nearly all women with patent Fallopian tubes retrograde menstruation is a common event; (2) we know that subsequently viable endometrium is present in the peritoneal cavity; (3) we know that menstrual endometrium is capable of growing in culture, has all the capacities to proliferate, has the necessary adhesion molecules to attach to the extracellular matrix in vitro, produces matrix metalloproteinases to allow for invading the extracellular matrix in vitro, and is able to induce angiogenesis necessary for its survival. What we do not know is if this sequence of events really happens in vivo. The theory of endometrial metaplasia of the peritoneal serosa may not be contradictory to the transplantation theory, but may be supplementary.

Etiology

Retrograde menstruation is a common event and a minor form of peritoneal endometriosis can be found in the majority of women laparoscoped irrespective of the reason for laparosocpy, be it subfertility, sterilisation or abdominal pain. However, in a subset of women the phenomenon endometriosis deteriorates and becomes an incapacitating disease.

It has been hypothesized that in most women there is an equilibrium between the development of minimal peritoneal lesions and the capacity of the abdominal cavity to resorb the endometrial tissue present in the abdominal cavity at the end of menstruation. This equilibrium is disturbed when larger amounts of tissue arrive in the abdominal cavity through (1) increased reflux secondary to an outflow obstruction, (2) increased reflux secondary to longer and heavier menstrual periods, or (3) increased peristalsis of the uterus.

Apart from the amount of endometrial tissue arriving in the abdominal cavity, the differences in the endometrium of women with as compared to women without endometriosis has gained much interest in recent years. A few examples: (1) Endometrium of women with endometriosis was shown to have a higher proliferative capacity and a higher potential of implantation and growth on peritoneal surfaces, (2) the enzyme aromatase cytochrome P450, essential for the synthesis of estrogen, has been identified in endometrium and in endometriotic implants of women with endometriosis and not in the endometrium of women without endometriosis. This implies that in endometriosis patients the endometrium itself creates its own stimulus. (3) VEGF protein as evaluated immunohistochemically showed a greater expression of VEGF in the late secretory phase in endometrium of women with endometriosis than in endometrium of controls.

The peritoneal environment in which the endometriotic lesion develops has been another area of research in recent years. All conceivable cytokines, macrophage products, hormones, prostaglandins, MMPs and TIMPs have been analysed in the peritoneal fluid, comparing women with and without endometriosis. Unfortunately, the results are lacking direction. Moreover, the studies were invariably performed in women that already had developed the disease. Therefore no firm conclusions can be drawn from this kind of studies regarding the etiology and pathogenesis of the endometriotic implant.

The Future

The etiology of endometriosis is multifactorial and therefore complex. New developments in biomedical research allow investigators to study gene and protein expression on a large scale. The research is moving away from hypothesis driven objectives and is partly evolving into a technology driven hunt. Genome and proteome wide screening methods may aid in elucidating the key pathways or players involved in the pathogenesis of diseases.

20

Baboon Model for Fundamental and Preclinical Research in Endometriosis

T.M. D'Hooghe[a,b], S. Debrock[b], Cleophas M. Kyama[a,b], D.C. Chai[a], S. Cuneo[a], J.A. Hill[c], J.M. Mwenda[a]

[a]Institute of Primate Research, Karen, Nairobi, Kenya; [b]Leuven University Fertility Center, Department of Obstetrics and Gynecology, University Hospital Gasthuisberg, Leuven, Belgium; [c]The Fertility Center of New England, Reading, MA, USA

Need for Primate Models in Endometriosis Research

Endometriosis is an important gynecological disease occurring in 10% of women between 16 and 50 years, and can be found in 50–70% of women with infertility and/or pain. Endometriosis is defined as the ectopic presence of endometrium outside the uterus, is diagnosed by laparoscopic inspection of the pelvis and by biopsy of endometriotic lesions with histological confirmation of the presence of endometrial glands and stroma. The presentation of endometriosis can vary from peritoneal disease (peritoneal vesicles, polyps, fibrotic areas of clear, white, red or black color), cystic ovarian endometriosis and deep rectovaginal disease. Adhesions related to endometriotic lesions are very common and may lead to tubal infertility. Endometriosis can be treated medically (hormonal suppression) or by surgery (excision of endometriotic lesions and adhesions), but recurrence is common, especially in cases of moderate to severe endometriosis. For many women, endometriosis has become a more or less chronic disease. Whereas a genetic basis for endometriosis is possible and environmental factors may play a role, many fundamental questions remain regarding pathogenesis of, spontaneous evolution of and subfertility associated with endometriosis. This can be explained because endometriosis only occurs in women and in nonhuman primates including baboons, rhesus monkeys, cynomolgus monkeys and De Brazza monkeys. In women, it is ethically impossible to perform controlled studies evaluating the fundamental questions mentioned above. In primate models, this can be ethically justified, since rodent models like mice or rabbits do not have spontaneous endometriosis and 'induced endometriosis' in these species looks very different from human endometriosis.

Comparative Advantages of Baboons vs. Other Monkeys

Compared to rhesus monkeys and cynomolgus monkeys, baboons have the following advantages:

1. Possibility to monitor the follicular and luteal phase by inspection of the perineum (inflation during follicular phase, deflation during luteal phase, ovulation occurring 2–3 days before deflation).
2. Continuous breeding (no seasonal breeding) in captivity.
3. Larger size (adult female: about 12–15 kg in captivity) and stronger health in captivity, allowing repeated blood samples and repeated surgical interventions like laparoscopies.
4. Accessibility of the cervix. The cervix can be cannulated in parous adult female baboons, allowing endometrial biopsies, flushing of the uterine cavity for embryo retrieval, minihysteroscopy.
5. Spontaneous endometriosis varying between minimal, mild, moderate and severe disease, using the adapted classification system of the American Society for Reproductive Medicine.
6. Induced endometriosis (after intrapelvic injection of menstrual endometrium) with similar macroscopic and microscopic presentation as spontaneous endometriosis.
7. Spontaneous peritoneal fluid. Peritoneal fluid is considered to be very important in the pathogenesis of endometriosis, and is very scant or absent in the other monkeys mentioned above. The baboon has about 2 ml of peritoneal fluid in the luteal phase.
8. Possibility to perform postcoital test. In fertility studies, it is important to ensure that coitus effectively occurred during the fertile period. During the late follicular phase, male baboons can be introduced in group cages or single cages with female baboons, but is not always sure whether coitus occurs, since it is practically impossible to monitor their behavior 24 h a day. In return for a banana, female baboons can be trained to present their perineal area, which allows investigators to take a vaginal swab, and examine the vaginal smear for the presence of sperm. This is very similar to the postcoital test performed in women during an infertility investigation. A positive test confirms that coitus has occurred.
9. In vivo culture system to study interaction between menstrual endometrium, peritoneal fluid, and pelvis and test new drugs that may inhibit this interaction.
10. Prevention studies: prevent the establishment of endometriosis by pretreatment of menstrual endometrium before intrapelvic injection (ex vivo treatment) or by intrapelvic injection of new drugs before intrapelvic seeding of menstrual endometrium, or by systemic injection of new drugs before intrapelvic seeding of menstrual endometrium.
11. Treatment studies: evaluate the effect of medical or surgical therapy in a controlled fashion in baboons with minimal, mild, moderate or severe endometriosis (spontaneous or induced).
12. Infertility studies: evaluate the effect of medical or surgical therapy on endometriosis-associated infertility with standardization for the degree of endometriosis (positive correlation between weight of menstrual endometrium used for intrapelvic injection and the stage of endometriosis), the presence of ovulation (perineal inspection), and the presence of coitus (behavioral observation, postcoital test).

Baboon Model for Research in Endometriosis

These advantages have become apparent after the baboon has been developed as a model for endometriosis research at the Institute of Primate Research in Nairobi, Kenya, since 1990, with support from the European Union, the Flemish Interuniversity Council (VLIR, Vlaamse Interuniversitaire Raad), Leuven University, Harvard Medical School (Fearing Research Laboratory) and the pharmaceutical industry.

This research in baboons has resulted in the following contributions in endometriosis research:

1. At IPR, where baboons of proven fertility in the wild are captured for research projects in reproduction, endometriosis occurs in about 20% of baboons, is mostly minimal to mild, and is significantly correlated with the number of previous hysterotomies.
2. The laparoscopic appearance and microscopic aspects of baboon endometriosis are similar to human endometriosis, but ovarian endometriosis was rarely seen. As in women, microscopic endometriosis in macroscopically normal peritoneum is rare.
3. The prevalence of endometriosis increases with the duration of captivity, probably due to the longer uninterrupted period of menstrual cycles with retrograde menstruation.
4. The score and stage of endometriosis increased in all baboons during follow-up over 2 years by laparascopic inspection every 6 months.
5. The cumulative incidence of histologically proven endometriosis in baboons with an initially normal pelvis was about 70% during follow-up over 3 years by laparoscopic inspection every 6 months.
6. High-dose immunosuppression did not have a major effect on the progression of endometriosis.
7. Moderate to severe endometriosis can be more successfully induced by intrapelvic seeding of menstrual endometrium than by intraperitoneal or subperitoneal injection of endometrium obtained during follicular or luteal phase.
8. Mild to moderate endometriosis is associated with a reduced monthly fecundity rate, whereas minimal endometriosis is associated with a normal monthly fecundity rate.
9. The reduced monthly fecundity rate in baboons with mild endometriosis may be related to an increased incidence and recurrence of the Luteinized Unruptured Follicle Syndrome.
10. Increased pelviperitoneal inflammation is present during menstruation and after intrapelvic injection of menstrual endometrium in baboons.
11. Subclinical pelvic inflammation occurs within the first week after a diagnostic laparoscopy in baboons.
12. Many antihuman mouse monoclonal antibodies for immune cells, cytokines and growth factors cross-react with baboons, as shown using FACS analysis and immunohistochemistry.

13 Baboons with endometriosis have a normal natural killer cell activity.
14 Treatment with TNF-α binding protein may prevent the development of endometrioiss and endometriosis-related adhesions.

The development of the baboon model for research in endometriosis has led the publication of more than 30 papers in international peer-reviewed journals, more than 10 chapters in international textbooks and has attracted the interest of the pharmaceutical industry to test new drugs in the prevention or treatment of endometriosis. The baboon model for endometriosis has become established and is accepted as the best animal model to study this disease. Research on this model has been awarded with the Endometriosis Prize at the World Congress of Fertility and Sterility in 1995, and with the General Program Prize at the Annual Meeting of the American Society for Reproductive Medicine in 2001. As a result, IPR has become a Center of Reference for primate research in endometriosis.

Further Reading

Awards

1 D'Hooghe TM, Bambra CS, Raeymaekers BM, Koninckx PR: Increased incidence and recurrence of recent corpus luteum without ovulation stigma (Luteinized Unruptured Foliccle-Syndrome?) in baboons (Papio anubis, Papio cynocephalus) with endometriosis. Presented at the 15th World Congress on Fertility and Sterility, Montpellier, France, September 17–22, 1995. Awarded the Endometriosis Award (TAKEDA).
2 D'Hooghe TM, Nugent N, Cuneo S, Chai D, Deer F, Debrock S, Mwenda J: Recombinant human TNF binding protein (r-hTBP-1) inhibits the development of endometriosis in baboons: a prospective, randomized, placebo- and drug-controlled study. Presented at the Annual Meeting of the American Society for Reproductive Medicine, Orlando, USA, October 22nd–24th 2001. Won one of the 2 General Program Prizes after nomination together with 9 other abstracts out of 1045 submitted abstracts. Fertil Steril 2001;76, O-2, S-1.

International peer-reviewed journals

1 D'Hooghe TM, Bambra CS, Cornillie FJ, Isahakia M, Koninckx PR: Prevalence and laparoscopic appearance of spontaneous endometriosis in the baboon (Papio Anubis, Papio Cynocephalus). Biol Reprod 1991;45:411–416.
2 Koninckx PR, D'Hooghe T, Oosterlynck D: Endometriosis: will the real natural history please stand up? (reply to letter). Fertil Steril 1991;56:590–591.
3 D'Hooghe TM, Bambra CS, Isahakia M, Koninckx PR: Evolution of minimal endometriosis in the baboon (Papio anubis, Papio cynocephalus) over a 12-month period. Fertil Steril 1992;58:409–412.
4 Cornillie FJ, D'Hooghe TM, Bambra CS, Lauweryns JM, Isahakia M, Koninckx PR: Morphological characteristics of spontaneous endometriosis in the baboon (Papio anubis, Papio Cynocephalus). Gynecol Obstet Invest 1992;34:225–228.
5 D'Hooghe TM, Bambra CS, Farah IO, Raeymaekers B, Koninckx PR: High intra-abdominal pressure during laparoscopy: effects on clinical parameters and lung pathology in baboons (Papio anubis, Papio cynocephalus). Am J Obstet Gynecol 1993;1969:1352–1356.
6 D'Hooghe TM, Bambra CS, Koninckx PR: Cycle fecundity in baboons of proven fertility with minimal endometriosis.Gynecol Obstet Invest 1994; 37;63–65.
7 D'Hooghe TM, Bambra CS, Suleman MA, Dunselman GA, Evers HL, Koninckx PR: Development of a model of retrograde menstruation in baboons (Papio anubis). Fertil Steril 1994;62:635–638.
8 Koninckx PR, Oosterlynck D, D'Hooghe TM, Meuleman C: Deeply infiltrating endometriosis is a disease whereas mild endometriosis could be considered a non-disease. Ann NY Acad Sci 1994;734:333–341.
9 D'Hooghe TM, Bambra CS, De Jonge I, Machai PN, Korir R, Koninckx PR: A serial section study of visually normal posterior pelvic peritoneum from baboons with and without spontaneous endometriosis. Fertil Steril 1995;63:1322–1325.
10 D'Hooghe TM, Bambra CS, Koninckx PR: Peritoneal fluid volume and steroid concentration in baboons with and without endometriosis. Arch Gynecol Obstet 1995;256:17–22.
11 D'Hooghe TM, Scheerlinck JP, Koninckx PR, Hill JA, Bambra CS: Deficient anti-endometrium lymphocyte mediated cytotoxicity but normal natural killler activity in baboons with endometriosis. Human Reprod 1995; 10:557–562.
12 D'Hooghe TM, Bambra CS, Raeymaekers BM, Hill JA, Koninckx PR: Immunosuppression can increase progression of spontaneous endometriosis in baboons. Fertil Steril 1995;64:172–178.
13 D'Hooghe TM, Hill JA: Natural Killer cell activity and endometriosis. Letter to the editor. Fertil Steril 1995;64:226–228.
14 D'Hooghe TM, Bambra CS, Raeymaekers BM, De Jonge I, Lauweryns JM, Koninckx PR: Intrapelvic injection of menstrual endometrium causes endometriosis in baboons (Papio cynocephalus, Papio anubis). Am J Obstet Gynecol 1995;173:125–134.
15 D'Hooghe TM, Bambra CS,De Jonge I, Lauweryns JM, Koninckx PR: The prevalence of spontaneous endometriosis in the baboon increases with the time spent in captivity. Acta Obstet Gynecol Scand 1996;75:98–101.
16 D'Hooghe TM, Bambra CS, Raeymaekers BM, Koninckx PR: Serial laparoscopies over 30 months show that endometriosis is a progressive disease in captive baboons (Papio anubis, Papio cynocephalus). Fertil Steril 1996; 65:645–649.
17 D'Hooghe TM, Bambra CS, Raeymaekers BM, Koninckx PR: Increased incidence and recurrence of recent corpus luteum without ovulation stigma (Luteinized Unruptured Follicle-Syndrome?) in baboons (Papio anubis, Papio cynocephalus) with endometriosis. J Soc Gynecol Invest 1996;3: 140–144.
18 D'Hooghe TM, Bambra CS, Raeymaekers BM, Koninckx PR: Disappearance of the ovulation stigma in baboons (Papio anubis, Papio cynocephalus) as determined by serial laparoscopies during the luteal phase. Fertil Steril 1996;65:1219–1223.
19 D'Hooghe TM, Bambra CS, Hill JA, Koninckx PR: Effect on endometriosis and the menstrual cycle on white blood cell subpopulations in the peripheral blood and peritoneal of baboons. Hum Reprod 1996;11:1736–1740.
20 D'Hooghe TM, Bambra CS, Raeymaekers BM, Koninckx PR: The cumulative incidence rate of endometriosis in baboons (Papio anubis, Papio cynocephalus) with an initially normal pelvis is 70% after 30 months. Obstet Gynecol 1996;88:462–466.
21 D'Hooghe TM, Bambra CS, Raeymaekers BM, Riday AM, Suleman MA, Koninckx PR: The cycle pregnancy rate is normal in baboons with stage I endometriosis but decreased in primates with stage II and stage III–IV disease. Fertil Steril 1996;66:809–813.
22 D'Hooghe TM, Bambra CS, Raeymaekers BM, Koninckx PR: Increased prevalence and recurrence of retrograde menstruation in baboons with spontaneous endometriosis. Human Reprod 1996;11:2022–2025.
23 Pijnenborg R, D'Hooghe T, Vercruysse L, Bambra C: Evaluation of trophoblast invasion in placental bed biopsies of the baboon, with immunohistochemical localization of cytokeratin, fibronectin and laminin. J Med Primatol 1996;25:272–281.
24 D'Hooghe TM, Bambra CS, Hill JA: A very sticky riddle. Fertil Steril 1997; 67:182–183.
25 D'Hooghe TM: Clinical relevance of the baboon as a model for the study of endometriosis. Fertil Steril 1997;68:613–625.
26 D'Hooghe TM, Bambra CS, De Jonge I, Lauweryns JM, Raeymaekers BM, Koninckx PR: Pregnancy does not affect endometriosis in baboons (Papio anubis, Papio cynocephalus). Arch Gynecol Obstet 1997;261:15–19.
27 D'Hooghe TM, Bambra CS, Raeymaekers BM, Hill JA: Pelvic inflammation induced by diagnostic laparoscopy in baboons. Fertil Steril 1999;72: 1134–1141.
28 D'Hooghe TM, Pudney J, Alves L, Peixe K, Hill JA: Immunobiology of the female reproductive tract in the baboon. Am J Primatol 2001;53:47–54.
29 D'Hooghe TM, Ling Xiao, Koninckx PR, Hill JA: Cytokine profile in peripheral blood and peritoneal fluid of women with deep and superficial endometriosis. Arch Gynecol Obstet 2001;265:40–44.

30 D'Hooghe TM, Bambra CS, Ling Xiao, Hill JA: The effect of menstruation and intrapelvic injection of endometrium on peritoneal fluid parameters in the baboon. Am J Obstet Gynecol 2001;184:917–925.
31 Pauwels A, Brouwer B, Cenijn P, Schepens P, D'Hooghe T, Delbeke L, Dhont M, De Sutter P, Weyler J: The risk of endometriosis associated with exposure to dioxin-like compounds: a case control study. Hum Reprod 2001;16:2050–2055.
32 D'Hooghe TM, Debrock S: Endometriosis, retrograde menstruation and peritoneal inflammation. Hum Reprod Update 2002;8:84–88.
33 Bergqvist A, D'Hooghe TM: The endometriosis enigma: cumulative retrograde menstruation, inflammation, disturbed egg quality and the debate on medico-surgical treatment of infertility. Hum Reprod Update 2002;8:65–69.
34 D'Hooghe TM: Endometriosis and subfertility: is the relationship resolved? Sem Reprod Med. March 2003, in press.
35 Devlieger R, D'Hooghe TM, Timmerman D: Uterine adenomyosis in the fertility clinic. Hum Reprod Update, accepted.
36 D'Hooghe TM, Mwenda JM, Debrock S: Immunomodulators and/or aromatase inhibitors: are they the next generation of treatment for endometriosis? Curr Opin Obstet Gynecol, in press.

Chapters in textbooks

1 D'Hooghe TM: Chapter 53. Natural history of endometriosis in baboons. In: Endometriosis today. Advances in research and practice. Proceedings of the Vth World Congress on Endometriosis. Edited by H. Minaguchi and O. Sugimoto. Parthenon Publishing, Canforth, pp 58–65.
2 D'Hooghe TM: Natural history of endometriosis in baboons: is endometriosis an intermittent and/or progressive disease? In: Endometriosis: Basic Research and Clinical Practice. Edited by Venturini L, Evers JLH. Parthenon Publishing, London, 1998;51–58.
3 Vercammen E, D'Hooghe TM, Hill JA: Endometriosis and recurrent miscarriage. Sem Reprod Med 2000;18:363–368.
4 D'Hooghe TM: Avances en el tratamiento farmacologico de la endometriosis (Advances in medical treatment of endometriosis). Reproducion Humana (Human Reproduction). Edited by Remohi J, Pellicer A, Simon C, Navarro J, IVI, Spain. Second Edition. McGraw-Hill, Interamericana de Espana, Madrid, 2002, pp 199–208.
5 D'Hooghe TM, Hill JA: Chapter 25. Endometriosis. In: Novak's Gynecology. Williams and Wilkins, 13th edition, 2002, pp 931–972.
6 D'Hooghe TM: Endometriosis. In: Novak's Gynecology: self-assessment and review. Lippincott, Williams and Wilkins, 2nd edition, 2002, pp 223–228.
7 D'Hooghe TM, Debrock S, Hill JA: Does retrograde menstruation exist? Critical analysis of the presence of endometrial cells in peritoneal fluid during menstrual and follicular phase. In: Endometriosis: Basic Research and Clinical Practice. Edited by Venturini L. Parthenon Publishing, London, 2002.
8 D'Hooghe TM, Mwenda JM: Baboon model for endometriosis. In: The baboon in Biomedical Research. VandeBerg JL, Tardif SD, Williams-Blangero S (eds). Kluwer Medical Publishers, in press.
9 D'Hooghe TM, Debrock S, Mwenda JM: Animal model for endometriosis. In: Endometriosis: Advances and Controversies. Tulandi T, Redwine D (eds). Marcel Dekker, NY, in press.
10 D'Hooghe TM, Debrock S, Mwenda JM: Animal models. In: Endometriosis in Clinical Practice. Edited by DL Olive. Martin Dunitz Publishing, London, in press.

21

Endometriosis in the Baboon

A.T. Fazleabas[a], A. Brudney[a], D. Chai[b], J. Mwenda[b]
[a]Department of Obstetrics and Gynecology, University of Illinois at Chicago, Chicago, IL, USA; [b]Institute for Primate Research, Nairobi, Kenya

Introduction

Endometriosis is an estrogen-dependent gynecological condition that affects an estimated five million American women during their reproductive life span. Although multiple theories exist regarding the etiology of the disease, the implantation hypothesis of Sampson is the most widely accepted. According to Sampson's hypothesis, fragments of menstrual endometrium are refluxed through the fallopian tubes into the peritoneal cavity, then attach and grow on peritoneal surfaces. However, the fundamental mechanisms by which menstrual endometrium adheres, proliferates and establishes a functional vasculature in an ectopic site remain to be elucidated. The baboon is a valuable and clinically relevant model to study the etiology and consequences to fertility of this enigmatic disease. Experimental evidence indicates that intrapelvic injections of menstrual endometrium can induce endometriosis in this primate, thereby supporting the basic tenets of Sampson's hypothesis. The experimentally induced baboon endometriosis model was first established by D'Hooghe et al., in a series of extensive studies at the Institute for Primate Research in Nairobi, Kenya. Intraperitoneal injection of menstrual endometrium resulted in endometriosis as defined by incidence of endometriotic lesions, histological criteria and cycle fecundity.

We propose that the establishment of endometriosis involves two distinct phases. The initial phase is invasive and dependent on ovarian steroids while the second phase is the active disease and capable of endogenous steroid production. We hypothesize that the marked increase in ERβ enables this estrogen-dependent disease to establish itself in the peritoneum. Estradiol acting via ERβ primarily can either directly or indirectly modulate proliferation, proteolysis and angiogenesis, all of which are associated with disease progression. This is Phase I of the disease and occurs during the first four to six months post inoculation. Once the disease is established, endogenous estradiol synthesis is initiated as a result of increasing aromatase activity and suppression of 17β hydroxysteroid dehydrogenase (HS) Type 2. This is Phase II of the disease which is evident between 6 and 10 months.

Evaluation of the Lesions

Endometriosis can take on a variety of appearances. A wide range of lesion types that were described in humans are also evident in the baboon model. For example, red raised nodules on the superficial peritoneum or reddish blue proliferative endometriotic nodules are most commonly encountered in the baboon. The white opaque peritoneal nodules that lack hemosiderin are also evident. The classic brownish focal adhesions are only observed at the latter stages of the disease. In the baboon, the majority of lesions that result following inoculation of menstrual endometrium are observed on the peritoneum and uterine surface.

Steroid Receptors

Endometriosis is associated with steroid hormones. Estrogen excess exacerbates endometriosis, while intervention therapies that inhibit estrogen production or action ameliorate the disease symptoms. Although endometriosis is an estrogen-dependent disease, endometriotic lesions have a different pattern of steroid hormone receptor expression than the eutopic endometrium. Simultaneous biopsies of ectopic and eutopic endometrium demonstrate differences in histochemical morphology and hormone receptor distribution. It has been reported that the concentration of ER and PR in endometriotic lesion homogenates were lower than in extracts from normal eutopic endometrium. In our baboon studies, imunocytochemical analyses for ERα, ERβ and PR revealed that ERβ was the dominant steroid receptor present in endometriotic explants regardless of the phase of the cycle. ERα and PR were present at lower levels compared to eutopic endometrium from the same baboon and did not show the predictable cyclic changes. Thus, it is evident from these studies that experimentally induced endometriosis is histologically similar to that obtained from women with spontaneous disease. Further, the presence of steroid receptors, particularly ERβ, suggests that these lesions can respond to both exogenous and endogenous estrogen at all phases of the menstrual cycle.

Endometriosis and Uterine Receptivity

Under the influence of ovarian steroids, the uterine endometrium undergoes profound modifications in cellular differentiation. In primates, at the appropriate phase of the menstrual cycle the uterus becomes 'receptive' and enables the blastocyst to attach. This 'receptive window' is initially dependent on estrogen and progesterone. However, further morphological and biochemical changes are induced within the uterus by signals from the developing embryo and following trophoblast invasion.

The endometrium of women with endometriosis is thought to be dysfunctional and may contribute to lowered fecundity in baboons; minimal endometriosis is not associated with infertility. However, more extensive peritoneal (moderate to severe on the human rating scale) is associated with subfertility following natural mating in the absence of any ovarian involvement. Alterations in endometrial development in patients with endometriosis may contribute to endometriosis-related infertility. Reports from several in-vitro fertilization embryo transfer programs indicate that patients with endometriosis have decreased implantation rates. Although histologically normal, examination of eutopic endometrium from women with endometriosis has revealed other defects. Ultrastructural defects have been reported in the endometrium of women with endometriosis. Molecular markers of endometrial receptivity are altered in patients with endometriosis; specifically, αvβ3 integrin expression patterns and HOX gene expression are aberrant in the endometrium of women with endometriosis.

Our results in the baboon demonstrate that two major biomarkers of uterine receptivity are markedly down-regulated in response to the infusion of chorionic gonadotrophin (CG) as early as one and four months post-inoculation. Glycodelin is a major secretory product of the glandular epithelium and α- smooth muscle actin (α-SMA) is induced in the sub-epithelial stromal cells and is a prerequisite for decidualization. In animals with induced disease, glycodelin expression as assessed by RT-PCR was non-detectable on day 8 post-ovulation and was markedly down-regulated, compared to controls, following CG infusion at days 10 and 14 post-ovulation. The decrease in mRNA was also correlated with the decrease or absence of protein. There was no detectable immunolocalization of glycodelin in the glandular epithelium of baboons with endometriosis following CG stimulation compared to controls. In contrast, the morphological appearance of the endometrium was normal in both the control animals and baboons with endometriosis. These data have also been recently substantiated in human studies. Microarray analysis of differentially regulated genes in women with and without endometrisois also reveals a significant downregulation of glycodelin expression in the endometrium of women with endometriosis during the window of receptivity.

The sub-epithelial stromal cells respond to CG by expressing α-SMA. Following CG infusion in baboons with endometriosis, α-SMA was not evident in the sub-epithelial stroma. However, the smooth muscle cells around the spiral arteries and the myometrium stained positively in both controls and baboons with endometriosis.

Thus, the baboon offers unique opportunities to study important aspects of human endometriosis, including pathogenesis, prevalence, incidence, spontaneous evolution, and endometriosis-associated subfertility. In women, it is impossible to perform the serial and controlled observational studies that are required for completely understanding this enigmatic disease. In addition, pelvic injection of menstrual endometrium offers an in vivo culture model for the study of interactions between the endometrium and peritoneum. Since this induction method can also lead to advanced stages of endometriosis it provides an attractive model to determine the basic physiology associated with the decrease in fertility.

22

The Future of Endometriosis Research: Genomics and Proteomics?

R.N. Taylor

Center for Reproductive Sciences, Department of Obstetrics, Gynecology and Reproductive Sciences, University of California, San Francisco, CA, USA

Introduction

The first phase of the human genome project, to generate a blueprint sequence of all 22 human autosomes and the two sex chromosomes, was completed in February 2001 [1]. As a result of this unprecedented international scientific effort, the nucleotide resolution of some ~30,000 genes in the human genome has been elucidated. The cost to identify and sequence the ~3 billion chemical bases was about USD 1 per nucleotide [2].

Even before this sequence information was finalized, biologists began applying new genetic information to complex human diseases for which heritable susceptibility was suspected. Among these is endometriosis, an enigmatic gynecological condition manifested by the growth of endometrial glands and stroma outside the uterine cavity. This condition may affect as many as 10–15% of women of reproductive age, however those having first-degree relatives with endometriosis have an even higher risk [3]. The etiology of endometriosis is complex and multifactorial [4]. While familial inheritance plays a role, multiple candidate genes appear to be involved [5, 6]. Because of this complexity, endometriosis is ideally suited as a target for genome-wide scanning. The genomics revolution has introduced an inductive paradigm to complement the traditional reductionist focus on hypothesis-driven biomedical research. With global and unbiased assessment of gene expression, seemingly random molecules can be integrated into functional pathways: so-called 'systems biology.' This experimental approach tends to generate, rather than test, hypotheses.

Genomics

In one of the first reports of endometriosis genomics, Giudice and colleagues [7] identified glycodelin-A (GdA) as a differentially expressed gene in the secretory endometrium of normal controls and women with endometriosis. The biology and action of GdA (previously referred to as progestagen-associated endometrial protein, endometrial alpha-2-globulin and PP14) are not fully understood, but this protein has been postulated to play an important role in embryonic receptivity of the endometrium [8]. Using Affymetrix® chips with microarrayed oligonucelotides, GdA mRNA concentrations were found to be 2- to 3-fold less in mid-secretory, eutopic endometrium of women with endometriosis compared to levels in mid-secretory, eutopic endometrium from normal subjects [9, 10]. These findings are consistent with proteomic analyses (described below) and a comprehensive meta-analysis showing reduced embryonic implantation rates in women with endometriosis undergoing in vitro fertilization [11]. A gene product that was found to be increased over 100-fold in secretory endometrium of women with endometriosis was the transcript encoding the enzyme carbonic anhydrase I [10].

Another powerful screening technique used in genomics research was introduced by Liang and Pardee in 1992 [12]. Differential display-polymerase chain reaction (dd-PCR), in a fashion analogous to subtraction hybridization, employs random sets of oligonucleotide primers and PCR to amplify fragments of differentially expressed cDNAs from different tissues. For these studies, endometriotic implants and biopsies of eutopic endometrium from normal and endometriosis cases were obtained in the mid-proliferative phase of the ovulatory cycle and analyzed. One interesting PCR product was observed to be consistently upregulated in ectopic and eutopic tissue of women with endometriosis relative to normal eutopic endometrium. When this cDNA fragment was cloned and sequenced it was found to correspond to a zinc-finger transcription factor called early growth response (EGR)-1 [9]. EGR-1 is an interesting candidate as it can be activated by cytokines and hormones that previously were associated with endometriosis, including IL-1beta, IL-6, TNF-alpha and estradiol. In addition, EGR-1 stimulates the expression of VEGF [13] and one of its receptors [14], providing a plausible link to enhanced angiogenesis in women with endometriosis [15] and its postulated role in this disease [16]. In a preliminary study, Yang and colleagues reported that EGR-1 protein is upregulated in the epithelial cells of endometriotic lesions [17].

Using cDNA libraries spotted onto 96-well nylon filters, Eyster et al. [18] found several immune-related gene products that were increased in endometriotic implants compared to normal endometrium. In particular, cDNAs encoding the immunoglobulin lambda-light chain and complement 1S proteins were discovered. Stromal cell structural protein genes (e.g., vimentin and alpha- and beta-actin) also were found to be overexpressed in endometriotic tissue. These observations are consistent with the immunologically activated state of endometriotic implants [4] and the fibrotic tendency of healing lesions [19].

Cytogenetics

Cytogenetic methods that assess gene content in intact or microdissected endometriotic tissues include chromosome satellite painting and comparative genomic hybridization. In the former case, Kosugi et al. [20] used fluorescence in situ hybridization (FISH) to find significantly increased aneuploidy and loss of heterozygosity of chromosome 17 in endometriotic lesions. Interestingly, this chromosomal region is the location of several tumor suppressor genes important in the normal regulation of the mitotic cell cycle. Among these are p53, which also appears to be lost in endometriosis [21] and the Tob-1 tumor suppressor, shown by Lebovic et al. [22] to be downregulated in endometriotic cells treated with IL-1beta.

Comparative genomic hybridization is a technique based on the competitive in situ hybridization of differentially labeled DNA from diseased and normal tissue to human metaphase spreads. Regions of loss of DNA sequences are seen as a decreased color ratio of the two fluorochromes used to label the respective DNAs. Using this method, Gogusev et al. identified several chromosomal regions in which gene copy numbers are altered in endometriosis, particularly loss of chromosome 1p and 22q [23]. These findings suggest that genomic instability is an inherent feature of endometriosis.

Proteomics

Although fewer examples have been reported, endometriosis research also is amenable to analysis by proteomic methods. These strategies identify, purify and sequence expressed proteins directly, rather than infer their expression via mRNA and cDNA intermediates. In cases where information from genomics and proteomics findings is complementary, this provides compelling evidence that the identified markers are likely to be significant and reproducible. Two such cases are presented as examples. The original observation by Sharpe et al. [24] that GdA protein biosynthesis was reduced in endometriosis tissue relative to unaffected, cycle-matched controls, was based on two-dimensional gel electrophoresis. Another proteomics example is the observation that circulating autoantibodies to carbonic anhydrase are more likely to be detected in women with endometriosis than controls [25]. Recently, Yeaman and colleagues determined that the antibodies in these cases recognize a common carbohydrate epitope [26]. Taken together with the recent genomics data from Kao et al. [10], one could speculate that overproduction of endometrial carbonic anhydrase in endometriosis might give rise to these autoantibodies.

Conclusions

The introduction of genomics and proteomics into the realm of endometriosis research will revolutionize the promise for gene-based diagnostic tests and rational development of genetically-targeted drugs. It is important, however, that gynecologists and scientists alike not lose sight of the fact that new data generated using these powerful technologies require critical analysis and verification. Despite the more rapid accumulation of knowledge afforded by these advances, the ethical application of new discoveries to our patients will continue to require painstaking care and consideration.

Acknowlegements

Supported by the NIH/NICHD, through cooperative agreements U54-HD31398, the NIH Office of Women's Health Research and U54-HD37321 as a part of the Special Cooperative Centers Program in Reproduction Research and by a research contract from the Women's Health Research Institute, Wyeth, Inc.

References

1 Lander ES, Linton LM, Birren B, et al: Initial sequencing and analysis of the human genome. Nature 2001;409:860–921.
2 Davies K: Cracking the Genome: Inside the Race to Unlock Human DNA. New York, Free Press, 2001.
3 Simpson JL, Elias S, Malinak LR, et al: Heritable aspects of endometriosis. I. Genetic studies. Am J Obstet Gynecol 1980;137:327–331.

4 Lebovic DI, Mueller MD, Taylor RN: Immunobiology of endometriosis. Fertil Steril 2001;75:1–10.
5 Bischoff FZ, Simpson JL: Heritability and molecular genetic studies of endometriosis. Hum Reprod Update 2000;6:37–44.
6 Stefansson H, Geirsson RT, Steinthorsdottir V, et al: Genetic factors contribute to the risk of developing endometriosis. Hum Reprod 2002;17:555–559.
7 Giudice LC, Kao LC, Yang J, et al: Global gene expression profiling of human endometrium from women with endometriosis using Genechip® microarray technology (abstract); in NICHD Conference on Endometriosis: Emerging Research and Intervention Strategies, Bethesda, April 2001, pp 62–63.
8 Mueller MD, Vigne J-L, Vaisse C, et al: Glycodelin: A pane in the implantation window. Semin Reprod Med 2000;18:289–298.
9 Taylor RN, Lundeen SG, Giudice LC: Emerging role of genomics in endometriosis research. Fertil Steril 2002;78:694–698.
10 Kao LC, Germeyer A, Tulac S, et al: Gene expression profiles in endometrium during the window of implantation differ in women with and without endometriosis. Endocrinology 2003, in press.
11 Barnhart K, Dunsmoor-Su R, Coutifaris C: Effect of endometriosis on in vitro fertilization. Fertil Steril 2002;77:1148–1155.
12 Liang P, Pardee AB: Differential display of eukaryotic messenger RNA by means of the polymerase chain reaction. Science 1992;257:967–971.
13 Yan SF, Fujita T, Lu J, et al: EGR-1, a master switch coordinating upregulation of divergent gene families underlying ischemic stress. Nat Med 2000;6:1355–1361.
14 Vidal F, Aragones J, Alfranca A, et al: Up-regulation of vascular endothelial growth factor receptor Flt-1 after endothelial denudation: role of transcription factor EGR-1. Blood 2000;95:3387–3395.
15 Tan XJ, Lang JH, Liu DY, et al: Expression of vascular endothelial growth factor and thrombospondin-1 mRNA in patients with endometriosis. Fertil Steril 2002;78:148–153.
16 Taylor RN, Lebovic DI, Mueller MD: Angiogenic factors in endometriosis. Ann NY Acad Sci 2002;955:89–100.
17 Yang S, Fang Z, Suzuki T, et al: Early growth response-1 (Egr-1) is selectively upregulated in epithelial cells of endometriosis: a mechanism for 17 beta-HSD-2 deficiency in endometriosis. J Soc Gynecol Invest 2003;10 (suppl):396 (abstract).
18 Eyster K, Boles A, Brannian J, et al: DNA microarray analysis of gene expression markers of endometriosis. Fertil Steril 2002;77:38–42.
19 Chegini N: Peritoneal molecular environment, adhesion formation and clinical implication. Front Biosci 2002;7:91–115.
20 Kosugi Y, Elias S, Malinak LR, et al: Increased heterogeneity of chromosome 17 aneuploidy in endometriosis. Am J Obstet Gynecol 1999;180:792–797.
21 Bischoff FZ, Heard M, Simpson JL: Somatic DNA alterations in endometriosis: high frequency of chromosome 17 and p53 loss in late-stage endometriosis. J Reprod Immunol 2002;55:49–64.
22 Lebovic DI, Baldocchi RA, Mueller MD, et al: Altered expression of a cell-cycle suppressor gene, Tob-1, in endometriotic cells by cDNA array analyses. Fertil Steril 2002;78:849–854.
23 Gogusev J, Bouquet de Joliniere J, Telvi L, et al: Detection of DNA copy number changes in human endometriosis by comparative genomic hybridization. Hum Genet 1999;105:444–451.
24 Sharpe KL, Zimmer RL, Griffin WT, et al: Polypeptides synthesized and released by human endometriosis differ from those of the uterine endometrium in cell and tissue explant culture. Fertil Steril 1993;60:839–851.
25 D'Cruz OJ, Wild RA, Haas GG Jr, et al: Antibodies to carbonic anhydrase in endometriosis: prevalence, specificity, and relationship to clinical and laboratory parameters. Fertil Steril 1996;66:547–556.
26 Yeaman GR, Collins JE, Lang GA: Autoantibody responses to carbohydrate epitopes in endometriosis. Ann NY Acad Sci 2002;955:174–182.

23

Quantitative Assessment of Endometrial-Peritoneal Interaction in vitro: A Non-Invasive Diagnostic Test for Women with Endometriosis?

S. Debrock[a], J.A. Hill[b], T.M. D'Hooghe[a]

[a]Leuven University Fertility Center, UZ Gasthuisberg, KU Leuven, Belgium; [b]The Fertility Center of New England, Reading, MA, USA

Introduction

According to the implantation theory [1] endometriosis arises as a result of retrograde menstruation of endometrial fragments through the fallopian tubes into the peritoneal cavity with subsequent implantation and growth of these endometrial cells onto the peritoneum. The exact mechanisms of the initial interaction between refluxed endometrium and peritoneum are still not completely understood.

Several in vivo and in vitro models have been developed to study these interactions between endometrium and peritoneum. D'Hooghe and coworkers [2] proposed the baboon as an in vivo culture model for the study of endometrial-periotoneal implantation. Studies in baboons indicated that the in vivo intrapelvic implantation potential of menstrual endometrium is higher than the implantation potential of endometrium derived from either the follicular or luteal phase [2]. Nude mice were used [3] to observe transplantation of human endometrium. Their experiments demonstrated that stromal cells are involved in the attachment process and glandular cells in the growth of endometriotic lesions [3]. In vitro, potential interactions have been studied using amnion membranes [4, 5] or biopsies of human peritoneum [6–8], but conclusions are controversial. One group suggested that an intact mesothelium may prevent endometrial fragments from adhering to the peritoneum [4, 5]. Electron microscopy supported this hypothesis that the mesothelial lining acts as a barrier against adhesion [8]. Using endometrial tissue isolated from antegradely shed menstrual effluent, this group observed that adhesion was exclusively seen at locations where the epithelium was damaged or absent [9]. In contrast, another group [6, 7] reported that endometrium can attach rapidly to the intact mesothelial surface of peritoneum. Confocal laser scanning microscopy and electron microscopy of peritoneal implants demonstrated an intact layer of viable mesothelial cells beneath sites of endometrial attachment. Both endometrial epithelium and stroma attached to the mesothelium. It appeared that endometrial adhesion to the mesothelium occurred within 1 h and that transmesothelial invasion occurred within 18–24 h [6, 7]. Using a coculture of endometrial and mesothelial cells and live cell imaging (confocal laser scanning microscopy) Witz and coworkers [10] showed that, after initial attachment, transmesothelial invasion of endometrial cells occurred within 6–12 h and that mesothelial regrowth or healing started within 24 h. Monolayers of human mesothelium were also used to evaluate the effects of cytokines on the adherence of endometrial stromal cells and whole endometrial fragments to mesothelium [11, 12]. Tumor necrosis factor-α (TNF-α) can promote the adhesion of human endometrial stromal cells to monolayered mesothelial cells [11]. Also interleukin-8 (IL-8) stimulated the adhesion of endometrial stromal cells to fibronectin in a dose-dependent way [13].

Quantitative Assay for the Endometrial-Peritoneal Interaction in vitro

Endometrial-Peritoneal Adhesion: The Phase of the Menstrual Cycle and the Stage of Endometriosis. We have recently investigated

[14] the effect of the phase of the menstrual cycle and the presence of absence of endometriosis on the adhesion between endometrium and peritoneum. We used mechanically fragmented endometrium in an in vitro model to determine whether endometrium could adhere to autologous peritoneal surfaces. Our results indicated that a high degree of endometrial-peritoneal adhesion (80–100%) occurred after culturing the explants for 48 h. Endometrial-peritoneal adhesion was observed in 88, 93, and 75% of the experiments performed with endometrium derived from menstrual, follicular or luteal phase endometrium, respectively. Our results indicated that endometrium obtained during menstrual, luteal or follicular phase had similar potential to implant on autologous peritoneum suggesting that the adhesion of endometrium onto peritoneum is a universal process, independent of the cycle phase. In our study, adhesion assays were performed with endometrium from patients with and without endometriosis including women with minimal, mild, moderate and severe endometriosis. Adhesion was observed in 89% of patients with a normal pelvis and in 78% of patients with endometriosis. No correlation was observed between endometrial-peritoneal adhesion and the stage of endometriosis as adhesion was found in 81% of patients with rAFS endometriosis stage I, 73% of patients with rAFS endometriosis stage II, 100% of patients with rAFS endometriosis stage III patients and 50% of patients with rAFS endometriosis stage IV disease.

Endometrial-Peritoneal Adhesion: Histology of Attachment and Implantation. Our histological results [14] suggest that a dynamic process occurs at the endometrial-peritoneal interface. Using light microscopy, a physical continuity between endometrial tissue and peritoneal biopsy was observed. Invasion of stromal cells into the peritoneum occurred rapidly with the formation of a multicellular layer under the mesothelium. In our experiments, the area of endometrium adherent to peritoneum consisted mainly of cells morphologically and immunohistochemically identified as stromal cells, suggesting that stromal cells may play an important role in in vivo attachment. Below the area of attachment of endometrial fragments, no identifiable mesothelium was present, and at the margins, a re-epithelialisation process rapidly integrated the implant into the peritoneal layer: mesothelial cells from the peritoneum started growing over the endometrial implant and the mesothelial epithelium continued into the glandular epithelium of the endometrial implant. Histological and structural changes were observed in and below the endometrial implants. Under the epithelialized layer, the endometrial stroma was replaced by a mixture of fine fibrillar and necrotic debris, whereas the glands remained clearly recognizable. Finally, in interstitial peritoneal tissue, well-structured collagen fibers disintegrated as a result of elastoid degeneration. Interestingly, these histological observations were made in several experiments using endometrium from different phases of the menstrual cycle, regardless to the presence or absence of endometriosis and the stage of endometriosis.

Quantitative Assay for the Endometrial-Peritoneal Interaction in vitro

In vitro models are limited by their descriptive and qualitative nature. A quantitative adhesion assay is needed to study in detail the interaction between peritoneum and endometrium. We have recently developed [15, 16] an in vitro assay for the quantitative assessment of the endometrial-peritoneal interaction using commercially available human mesothelial cells (CRL-9444) and human endometrial epithelial carcinoma cells (CRL-1671). Adhesion of [^{35}S]methionine-labelled human endometrial cells to human mesothelial cells was measured in a quantitative way. Binding occurred rapidly within 1 h of incubation in a time-dependent manner. We also evaluated the effect of the cytokines TNF-α, IL-6 and IL-8 on the adhesion between mesothelial and endometrial cells. The in vitro adhesion of labelled endometrial cells to mesothelial cells, pretreated with different concentrations (0–1,000 U/ml) of TNF-α, IL-6 or IL-8, was significantly inhibited by the different cytokines in a dose-dependent way.

Conclusion

Early adhesion between endometrial cells and pelvic peritoneum is accepted to be a key process in the development of endometriosis. This process has been studied using an in vitro approach. A qualitative in vitro adhesion model, using fragmented endometrium and peritoneal biopsies, mimicks the in vivo situation in women, but has several shortcomings in comparison with the in vivo situation. Firstly, endometrial tissue and cells obtained after a transcervical uterine biopsy are probably different from endometrial tissue and cells present in the peritoneal fluid in vivo. Secondly, the collection, handling and culture of peritoneum may alter the propensity of peritoneum to adhere to endometrium. Therefore, short-term in vivo coculture of menstrual endometrium, peritoneal fluid and peritoneum is needed and being developed in the baboon [17] to study the early endometrial-peritoneal interactions in a clinically and biologically meaningful way. A quantitative in vitro adhesion assay has been developed to determine inhibition and/or stimulation of endometrial-mesothelial adhesion by several cytokines, growth factors and proteinases. Using autologous endometrial cells, this assay may also be developed as a potential screening test for women at risk for the development of (recurrent) endometriosis. Moreover, this model may be used to develop new therapeutic approaches for the treatment of endometriosis. These in vitro and in vivo adhesion assays will allow to study the mechanisms involved in endometrial-peritoneal implantation and provide insight in the pathogenesis of endometriosis.

References

1 Sampson J: Peritoneal endometriosis due to the menstrual dissemination of endometrial tissue into the peritoneal cavity. Am J Obstet Gynecol 1927; 14:422–469.
2 D'Hooghe TM, Bambra CS, Raeymaekers BM, De Jonge I, Lauweryns JM, Koninckx PR: Intrapelvic injection of menstrual endometrium causes endometriosis in baboons *(Papio cynocephalus, Papio anubis)*. Am J Obstet Gynecol 1995;173:125–134.
3 Nisolle M, Casanas-Roux F, Donnez J: Early-stage endometriosis: adhesion and growth of human menstrual endometrium in nude mice. Fertil Steril 2000;74:306–312.
4 Van der Linden PJ, de Goeij AF, Dunselman GA, Erkens HW, Evers JL: Endometrial cell adhesion in an in vitro model using intact amniotic membranes. Fertil Steril 1996;5:76–80.
5 Groothuis PG, Koks CA, de Goeij AF, Dunselman GA, Arends JW, Evers JL: Adhesion of human endometrium to the epithelial lining and extracellular matrix of amnion in vitro: an electron microscopic study. Hum Reprod 1998;13:2275–2281.
6 Witz CA, Montoya-Rodriguez AI, Schenken RS: Whole peritoneal explants: a novel model of the early endometriosis lesion. Fertil Steril 1999; 71:56–60.
7 Witz CA, Thomas MR, Montoya-Rodriguez AI, Nair AS, Centonze VE, Schenken RS: Short-term culture of peritoneum explants confirms attachment of endometrium to intact peritoneal mesothelium. Fertil Steril 2001; 75:385–390.
8 Groothuis PG, Koks CA, de Goeij AF, Dunselman GA, Arends JW, Evers JL: Adhesion of human endometrial fragments to peritoneum in vitro. Fertil Steril 1999;71:1119–1124.

9 Koks CA, Groothuis PG, Dunselman GA, de Goeij AF, Evers JL: Adhesion of shed menstrual tissue in an in-vitro model using manion and peritoneum: a light and electron microscopic study. Hum Reprod 1999;14:816–822.
10 Witz CA, Cho S, Centonze VE, Montoya-Rodriguez IA, Schenken RS: Time series analysis of transmesothelial invasion by endometrial stromal and epithelial cells using three-dimensional confocal microscopy. Annual Meeting of the American Society for Reproductive Medicine, San Diego, October 21–25, 2000.
11 Zhang R, Wild RA, Ojago JM: Effect of tumor necrosis factor-α on adhesion of human endometrial stromal cells to peritoneal mesothelial cells: an in vitro system. Fertil Steril 1993;59:1196–1201.
12 Wild RA, Zhang R, Medders D: Whole endometrial fragments form characteristics of in vivo endometriosis in a mesothelial cell co-culture system: an in vitro model of the histogenesis of endometriosis. J Soc Gynecol Invent 1994;1:165–168.
13 Garcia-Velasco JA, Aydin A: Interleulin-8 stimulates the adhesion of endometrial stromal cells to fibronectin. Fertil Steril 1999;72:336–340.
14 Debrock S, Vanderperre S, Meuleman C, Moerman Ph, Hill JA, D'Hooghe TM: In vitro adhesion of endometrium to autologous peritoneal membranes: effect of the cycle phase and the stage of endometriosis. Hum Reprod 2002;17:2523–2528.
15 Debrock S, Destrooper B, Vander Perre S, Hill JA, D'Hooghe TM: Tumor necrosis factor-alpha inhibits the in vitro adhesion of endometrial epithelial cells to mesothelial cells. Oral presentation at the 8th Biennial World Congress on Endometriosis, San Diego, USA, February 2002. Fertil Steril 2002;77(suppl 1):S15.
16 Debrock S, Destrooper B, Vander Perre S, Hill JA, D'Hooghe TM. Effect of interleukin-6 and interleukin-8 on the in vitro adhesion of endometrial epithelial cells to mesothelial cells. Oral presentation at the 58th Annual Meeting of the American Society for Reproductive Medicine, October 12–17, 2002, Seattle, USA. Fertil Steril 2002;76(suppl):O-233.
17 D'Hooghe TM: Clinical relevance of the baboon as a model for the study of endometriosis. Fertil Steril 1997;68:613–625.

24

Medical Treatment of Endometriosis

C. Simón, J. García-Velasco

Instituto Valenciano de Infertilidad, Valencia University, Valencia, Spain

Considering that the pathogenesis of this disease is still unknown, it is not easy to delineate clear lines of treatment, whether surgical or medical. In patients with endometriosis, we only treat symptoms: pelvic pain or infertility. Most of the different treatments assayed have never been compared in appropriate randomized, controlled trials. As a result, most of the experience comes from empiric knowledge. During this lecture, we will review the different options when trying to choose a medical treatment for a patient with endometriosis.

Endometriosis has a well known hormonal dependence, and ectopic tissue depends on steroids to grow and provoque symptoms. In a similar fashion to surgical ablation of endometriotic lesions, current medical options have been based on hormonal suppression. A reduction in the activity of the disease has been shown to decrease pain associated with this disease; improvement in fertility through medical treatment is still a matter of debate.

Classical Approaches

Pseudopregnancy/Progestins. Endometriosis has been shown to improve with pregnancy, so different progestin protocols have been proposed. Most authors have shown objective signs of improvement, although their main side effect is a not so infrequent spotting or bleeding, inducing a high rate of treatment drop-out.

Oral Contraceptives. They have been the gold standard in the cronic treatment of endometriosis; the reduction in menstrual flow not only diminished dismenorrhea but also diminishes the risk by reducing the chance of retrograde menstruation. New oral contraceptives with much lower estradiol concentrations are now used continuously to prevent progression of the disease. However, not all authors agree on their beneficial effect.

Danazol. This derivate from 17-α-ethinyl testosterone induces a hypoestrogenic state by suppressing pituitary gonadotropins and inhibiting various steroidogenesis enzymes. It binds to progesterone receptor, inducing also some of the effects observed with progestins. Danazol also has affinity for the androgen receptor, and most of the secondary effects are due to an increase in testosterone (acne, weight gain, lowering of the voice, changes in the hair and skin ...). In a recent report, this drug has been linked to a higher risk of ovarian cancer in women who were treated with danazol.

GnRH Analogues. These synthetic derivates from the GnRH decapeptide are one of the most common treatments used nowadays. All authors agree that GnRH analogues are useful in improving pelvic pain symptoms in women with endometriosis. Depot formulations allow a single injection every month, or even once every 3 months, achieving a great compliance. They induce a hypoestrogenic state, the key of their success but also the reason for the concerns. When treatment is prolonged, loss of bone mineral density may occur. This problem may be solved by adding small amounts of estrogen – add back therapy – to reduce this risk.

New Therapeutic Approaches

Aromatase Inhibitors. Endometriotic tissue as well as eutopic endometrium from women with endometriosis express aromatase P450, a feature not observed in eutopic endometrium without the disease. This could be one of the mechanisms by which there is an excess of estrogen production that may stimulate tissue growth. Aromatase inhibitors, although not a risk-free treatment, have been successfully assayed in recurrent, difficult cases of women with endometriosis.

Antiprogestins. They have the capacity to atrophy both the endometrium and ectopic tissue, with a reduction in the mitotic activity of the endometrial glands. The unopposed estrogen action does not seem to be a problem. Data available from rhesus monkeys seems promising, but no data is still available in humans.

SPRMs (Selective Progesterone Receptor Modulators). These new compounds have both agonist and antagonist activities depending upong the site of action. Basic and animal experimental data is showing very attractive results. Human experimentation is still at phase I–phase II, and although they will reach the market by 2005–2006, SPRM offer a completely different and new approach to the medical treatment of this disease.

GnRH Antagonists. These new drugs directly block GnRH receptor in the pituitary, inducing a hypoestrogenic state similar to GnRH agonists but in a shorter period of time and without the flare-up effect. In a recent report, weekly administration of cetrorelix 3 mg has shown promising results without hypoestrogenic symptoms, as estradiol levels oscillated around 50 pg/ml.

References

1 Lessey BA: Medical management of endometriosis and infertility. Fertil Steril 2000;73:1089–1096.

2 Barbieri RL: Hormone treatment of endometriosis: the estrogen threshold hypothesis. Am J Obstet Gynecol 1992;166:740–745.
3 Hughes EG, Fedorkow DM, Collins JA: A quantitative overview of controlled trials in endometriosis-associated infertility. Fertil Steril 1993;59: 963–970.
4 Weideman M: Danazol linked to ovarian cancer. Lancet Oncol 2002;3: 261.
5 Olive DL: Role of progesterone antagonists and new selective progesterone receptor modulators in reproductive health. Obstet Gynecol Surv 2002;57: S55–S63.
6 Felberbaum R, Kuepker W, Diedrich K: GnRH-antagonists: new pathways in the treatment of endometriosis. Gynecol Endocrinol 2003;17(suppl 1):27.

25

Inhibition of Endometrial Peritoneal Attachment in the Prevention and Treatment of Endometriosis

P.G. Groothuis, G.A.J. Dunselman

Department of Obstetrics and Gynaecology, Research Institute GROW, Maastricht University, Maastricht, The Netherlands

During the reproductive period in premenstrual women menstrual effluent containing shed endometrial tissue and blood is disposed off vaginally by the body each month. Part of this effluent is pushed in the opposite direction through the fallopian tubes and reaches the abdominal cavity. According to Sampson, endometrial tissue implants in the peritoneal surface and develops into peritoneal endometriosis. At laparoscopy active superficial endometriotic lesions are easily recognized by the abundance of vasculature.

Histology usually demonstrates that these implants consist of foci with functional endometrial and stromal cells covered by mesothelium.

Longer menstrual periods and heavier blood flow, as well as congenital malformations leading to obstruction of antegrade blood flow, increase the likelihood of developing endometriosis. Therefore, avoiding the arrival of endometrial tissue in the abdominal cavity would be a first option to prevent the development of peritoneal endometriosis. This may be achieved either by hysterectomy or tubal occlusion. Although these options may seem logical, these are not acceptable choices for prophylaxis. Reducing the amount or composition of menstrual endometrium by hormonal treatment may also reduce the risk of developing endometriosis.

An alternative way to prevent the development of peritoneal endometriosis is to minimize the adherence of the monthly arriving endometrial cells or tissue to the peritoneal lining. Tissue adhering to, or actively invading the peritoneal lining has never been identified, which suggests that the implantation of endometrial tissue is a very rapid process. Indeed, in the chick chorioallantois membrane, a model for endometriosis, we observed that human endometrium has the capacity to invade and form endometriosis-like lesions within three days after transplantation. This urges for clarity concerning the interactions that occur between the mesothelium and the regurgitated endometrium upon arrival in the abdominal cavity.

The abdominal cavity is actually not a cavity, but a compartment filled with fluid. It is interesting to note that during menstruation only a few milliliters of peritoneal fluid can be recovered from the pouch of Douglas during laparoscopy. This indicates that the surfaces of the various intra-abdominal structures are separated by a fluid film. This leads to an intimate contact of the regurgitated menstrual endometrium with the mesothelial cells lining the peritoneum. The classic functions attributed to the mesothelium are to create a frictionless interface for the free movement of apposing organs and tissue and to provide a protective barrier. However, the mesothelium has also been implicated in various other processes such as fluid and cell transport, initiation and resolution of inflammation, tissue repair, lysis of fibrin deposits preventing adhesion formation, protection against microorganisms and, possibly, tumour dissemination [1].

Whether the mesothelium is receptive for the endometrium cells or acts as a barrier [2] is still an issue of debate. We have previously shown in vitro that endometrial cells preferably attach to amnion and peritoneum in areas where the epithelial lining was absent or damaged. Based on our findings we postulated that an intact epithelial lining prevents the development of peritoneal endometriosis. However, other investigators showed that endometrial cells are able to attach to the mesothelial cells, through interaction of CD44 and hyaluronic acid (HA), which suggests that the mesothelium is receptive as well [3]. The binding of various types of disseminating tumor cells to the mesothelium is also mediated by the interaction of CD44 expressed on the cells and HA present on the mesothelial cell surface. It appears therefore, that this is a general event that precedes implantation of many types of ectopic cells present in the abdominal cavity. It has also been shown that β1-integrins expressed on ectopic cells are involved in the adhesion process, suggesting that interaction with exposed submesothelial surfaces also occur in vivo. Furthermore, it was shown that cells expressing CD44 that bind HA start to express hyaluronidase in order to degradate the HA. Hyaluronidase is present in endometrial cells as well as in macrophages. Apparently, it does not matter whether the HA coat is present on the mesothelium or not, the endometrium has the ability to deal with both situations. Regurgitated menstrual blood contains many leukocytes, however, and evokes an inflammatory response when arriving in the peritoneal cavity, which complicates matters.

Since cells can adhere to the peritoneum irrespective of the presence of HA, the solution has to be found at the level of the endometrium. When attempting to prevent binding of endometrial cells to the peritoneal surface, CD44 and β1-integrins on the surface of endometrial cells are the likely targets. There are three options: (1) reduce CD44 and β1-integrin expression during the menstrual cycle prior shedding, (2) reduce CD44 and β1-integrin expression intraabdominally after shedding, or (3) mask CD44 and β1-integrins on endometrial cells present in the abdominal cavity. A compound that reduces CD44 and β1-integrin expression just prior to menstruation may aid to minimize the chance of endometrial cell adhesion after regurgitation. Recently, investigators have tested whether mesothelial cells can be used to deliver a protein in vivo. Mesothelial cells were isolated and transfected in vitro with an expression vector containing the cDNA of the protein of interest. These cells started to produce this protein in vitro. The mesothelial cells were injected intraabdominally and after a few days the protein could be measured in the peripheral blood, which proves that the mesothelial cells were able to mix among the other mesothelial cells and synthesize the protein [4]. This opens an avenue of new possibilities for the prevention of adhesion of ectopic cells.

References

1 Mutsaers SE: Mesothelial cells: Their structure, function and role in serosal repair. Respirology 2002;7:171.
2 Dunselman GAJ, et al: The mesothelium, Teflon or Velcro. Hum Reprod 2001;16:605.
3 Dechaud H, et al: Mesothelial cell-associated hyaluronic acid promotes adhesion of endometrial cells to mesothelium. Fertil Steril 2001;76:1012.
4 Devin CJ, et al: Pleural space as a site of ectopic gene delivery? Chest 2003; 123:202.

26

Steroid and Cytokine Regulation of Matrix Metalloproteinases and the Pathophysiology of Endometriosis

K.G. Osteen[a], T.M. Igarashi[b], G.R. Yeaman[a], K.L. Bruner-Tran[a]

[a]Women's Reproductive Health Research Center, Department of Obstetrics and Gynecology, Vanderbilt University School of Medicine, Nashville, TN, USA; [b]Teikyo University Hospital, Ichihara Branch, Department of Obstetrics and Gynecology, Ichihara City, Chiba, Japan

The endometrium of humans and other primates is a unique adult organ system that undergoes cyclic breakdown and regrowth throughout a women's reproductive life. Endometriosis, defined most simply as the ectopic growth of endometrial glandular and stromal elements outside the uterine corpus, appears to represent an untoward consequence of menstruation. More than 75 years ago, Sampson theorized that the development of endometriosis can be linked to the mechanical transfer of endometrial tissue to the peritoneal cavity via retrograde menstruation [1]. Endometriosis is classified as a benign condition even though the establishment of ectopic growth is an invasive event that requires degradation of extracellular matrix (ECM) proteins through the action of matrix-degrading enzymes. The cell-specific expression of numerous matrix metalloproteinases (MMPs) has been documented during the extensive turnover of endometrial tissue that occurs in response to changing patterns of ovarian steroid production during each menstrual cycle [2]. Steroid-mediated expression and action of MMPs during the menstrual cycle may provide an important mechanistic link between Sampson's original theory of retrograde menstruation and the invasive processes necessary for ectopic establishment of endometriosis [1, 3–6].

Endometriosis is often linked to infertility and investigators have begun to focus more attention on the possibility that the eutopic endometrium of women who suffer from endometriosis may hold clues to the disease process. The chance of reproductive success during each menstrual cycle is dependent upon appropriate cell-specific responses to steroids, including regulation of MMP expression. In this regard, we have recently identified an altered pattern of endometrial MMP expression among women with endometriosis during the 'window of implantation.' Our studies revealed that endometrial expression of MMP-3, MMP-7 and MMP-11 mRNA remains elevated during the progesterone-dominated secretory phase in women with endometriosis, a period when these MMPs are normally suppressed [7, 8]. Our studies further indicated that the altered expression of MMPs found in women with endometriosis leads to an increased capacity of their eutopic endometrial tissue to establish ectopic lesions in a nude mouse model of the disease [7, 8]. These recent observations continue to support previous studies that suggest that defects in progesterone sensitivity among women with endometriosis may be linked to the pathophysiology of the disease [9, 10].

Normal regulation of the endometrial MMP system in response to changing patterns of ovarian estrogen or progesterone production requires local cell-specific interactions that are mediated by a wide variety of growth factors and cytokines [2]. For example, retinoic acid (RA), the biologically active metabolite of Vitamin A, is synthesized in the endometrium during the process of stromal cell decidualization [11, 12]. Treatment of human endometrium in vitro with RA can mimic many of the effects of progesterone during the process of endometrial maturation, including the suppression of cell-specific MMP expression [12]. Not unexpectedly, we have shown that progesterone-mediated suppression of MMPs during preparation for pregnancy requires the secondary action of RA, acting in concert with an additional paracrine mediator, transforming growth factor-beta (TGF-β) [13]. Furthermore, ovarian progesterone and locally produced RA appear to act in concert to regulate an increase in the expression of TGF-β2 expression during normal preparation for pregnancy [11, 14]. In women with endometriosis, induction of appropriate paracrine signals during preparation for pregnancy appears to be diminished, leading to a failure to suppress MMP gene and protein expression. Our studies support this position since the failure of progesterone to suppress MMP-3 and MMP-7 in endometrial organ cultures established from women with endometriosis can be 'corrected' by supplementing progesterone treatment with RA and TGF-β [7].

Although, numerous studies now support the position that the endometrium of women with endometriosis lacks a normal response to progesterone, the cellular pathophysiology of progesterone failure among these patients is not completely understood. In this regard, our studies suggest that local pro-inflammatory cytokine production by epithelial cells or immune cells may influence MMP regulation in the endometrium by affecting stromal cell responsiveness to progesterone [15]. Both decreased stromal responses to progesterone and increased stromal responses to local pro-inflammatory cytokine production may contribute to alterations in MMP regulation associated with endometriosis. For example, we have recently shown that in vitro treatments with the pro-inflammatory cytokine interleukin-1 (IL-1) alters progesterone receptor expression (PR-A/PR-B) in isolated endometrial stromal cells, resulting in an increased expression of MMP-3 [16]. Since endometrial stromal cells can mediate epithelial cell-specific MMP-7 expression in the endometrium [14, 17], alterations in PR-A/PR-B ratio in stromal cells would disrupt both stromal-specific and epithelial-specific MMP expression. In vivo studies of the endometrium acquired from endometriosis patients reveals similar alterations in the PR-A/PR-B ratio in eutopic endometrium and others have noted an absence of PR-B expression at ectopic sites [16, 18]. We have previously demonstrated that endometrial epithelial cells can produce IL-1α and this cytokine can stimulate MMP-3 expression by stromal cells in the absence of adequate progesterone support [15]. Interestingly, stromal cells isolated from the endometrium of women with endometriosis and cultured in the presence of epithelial cell-conditioned media become less responsive to progesterone, which indicates that a defect in epithelial to stromal cell communication may alter MMP regulation in these patients [19].

In addition to the above studies, our recent work with dioxin, an environmental toxin associated with endometriosis, further suggests that disruption of cellular communication may promote development of this disease. We found that in vitro treatments of endometrium from normal donors with dioxin results in an alteration of the PR-B/PR-A ratio that mimics tissue acquired from endometriosis patients. More specifically, we found that in vitro treatments with dioxin activates an epithelial-mediated pathway involving proinflammatory cytokines that disrupts stromal cell responses to progesterone [16]. Interestingly, in women and non-human primates, dioxin exposure prior to or during pregnancy is associated with an increased rate of spontaneous abortion [20–22]. Whether or not dioxin-mediated disruption of endometrial MMP expression contributes to the pathophysiology of endometriosis remains to be determined. Nevertheless, the complex and interactive system required for normal endometrial MMP regulation in response to steroids provides a remarkable biological balance. In the absence of this balance, the buffering system fails to prevent the overexpression of these enzymes, which is associated with numerous diseases. Our studies suggest that the normal interactive role of steroids and locally produced factors is disrupted among women with endometriosis, which may be related to the altered patterns of MMP expression we have recently observed to the pathophysiology of this disease [7].

Acknowledgements

Supported by NICHHD/NIH grants and through cooperative agreement as part of the Specialized Cooperative Centers Program in Reproductive Research (U54-HD37321). Also supported by the US Environmental Protection Agency and the Endometriosis Association.

References

1 Sampson JA: Peritoneal endometriosis due to menstrual dissemination of endometrial tissue into the peritoneal cavity. Am J Obstet Gynecol 1927; 14:422–469.
2 Curry T, Osteen KG: The ovarian and uterine matrix metalloproteinase system: Changes, regulation and impact throughout the reproductive cycle. Endocrine Reviews (in press).
3 Spuijbroek MD, Dunselman GA, Menheere PP, Evers JL: Early endometriosis invades the extracellular matrix. Fertil Steril 1992;58:929–33.
4 Saito T, Mizumoto H, Kuroki K, Fujii M, Mori S, Kudo R: Expression of MMP-3 and TIMP-1 in the endometriosis and the influence of danazol. Nippon Sanka Fujinka Gakkai Zasshi [in Japanese] 1995;47:495–496.
5 Osteen KG, Bruner KL, Sharpe-Timms KL: Steroid and growth factor regulation of matrix metalloproteinase expression and endometriosis. Semin Reprod Endocrinol 1996;14:247–255.
6 Bruner KL, Matrisian LM, Rodgers WH, Gorstein F, Osteen KG: Progesterone suppression of matrix metalloproteinases prevents establishment of endometriotic-like human lesions in nude mice. J Clin Invest 1997;99: 2851–2857.
7 Bruner-Tran KL, Eisenberg E, Yeaman GR, Anderson TA, McBean J, Osteen KG: Steroid and cytokine regulation of matrix metalloproteinase expression in endometriosis and the establishment of experimental endometriosis in nude mice. J Clin Endocrinol Metab 2002;87:4782–4791.
8 Bruner-Tran KL, Webster-Clair D, Osteen KG: Experimental endometriosis: the nude mouse as a xenographic host (review). Ann NY Acad Sci 2002; 955:328–339; discussion 340–342, 396–406.
9 Lessey BA, Metzger DA, Haney AF, McCarty KS, Jr: Immunohistochemical analysis of estrogen and progesterone receptors in endometriosis: comparison with normal endometrium during the menstrual cycle and the effect of medical therapy. Fertil Steril 1989;51:409–415.
10 Bergqvist A, Ferno M: Oestrogen and progesterone receptors in endometriotic tissue and endometrium: comparison of different cycle phases and ages. Hum Reprod 1993;8:2211–2217.
11 Osteen KG, Keller NR, Feltus FA, Melner MH: Paracrine regulation of matrix metalloproteinase expression in the normal human endometrium (review). Gynecol Obstet Invest 1999;48(suppl 1):2–13.
12 Osteen KG, Bruner-Tran KL, Ong D, Eisenberg E: Paracrine mediators of endometrial matrix metalloproteinase expression: potential targets for progestin-based treatment of endometriosis (review). Ann NY Acad Sci 2002 955:139–146; discussion 157–158, 396–406.
13 Osteen KG, Igarashi TM, Bruner-Tran KL: Progesterone action in the human endometrium: induction of a unique tissue environment which limits matrix metalloproteinase (MMP) expression. Front Biosci 2003;1;8: D78–D86.
14 Bruner KL, Rodgers WH, Gold LI, Korc M, Hargrove JT, Matrisian LM, Osteen KG: Transforming growth factor beta mediates the progesterone suppression of an epithelial metalloproteinase by adjacent stroma in the human endometrium. Proc Natl Acad Sci USA 1995;92:7362–7366.
15 Keller NR, Sierra-Rivera E, Eisenberg E, Osteen KG: Progesterone exposure prevents matrix metalloproteinase-3 (MMP-3) stimulation by interleukin-1alpha in human endometrial stromal cells. J Clin Endocrinol Metab 2000;85:1611–1619.
16 Igarashi TI, Bruner-Tran KL, Edwards D, Eisenberg E, Osteen KG: Differential regulation of progesterone receptor (PR) by TCDD in isolated human endometrial stromal cells versus stromal-epithelial co-cultures. 58th Annual Meeting of the American Society of Reproductive Medicine, Seattle WA, 2002.
17 Osteen KG, Rodgers WH, Gaire M, Hargrove JT, Gorstein F, Matrisian LM: Stromal-epithelial interaction mediates steroidal regulation of metalloproteinase expression in human endometrium. Proc Natl Acad Sci USA 1994;11;91:10129–33.
18 Attia GR, Zeitoun K, Edwards D, Johns A, Carr BR, Bulun SE: Progesterone receptor isoform A but not B is expressed in endometriosis. J Clin Endocrinol Metab 2000;85:2897–2902.
19 Bruner-Tran KL, Yoder C, Eisenberg E, Osteen KG: Altered stromal-epithelial cell communication may contribute to reduced progesterone sensitivity in endometriosis. 58th Annual Meeting of the American Society of Reproductive Medicine, Seattle WA, 2002.
20 Damewood M: The association of endometriosis and repetitive (early) spontaneous abortions. Semin Reprod Endocrinol 1989;7:155–160.
21 Bowman RE, Schantz SL, Weerasinghe NCA, Gross M, Barsotti D: Chronic dietary intake of 2,3,7,8-tetrachlorodibenzo-*p*-dioxin (TCDD) at 5 or 25 ppt in the monkey: TCDD kinetics and dose-effect estimates of reproductive toxicity. Chemosphere 1989;18:243–252.
22 Sweeney A: Reproductive epidemiology of dioxins; in Schecter A (ed): Dioxins and Health. New York, Plenum Press, 1994, pp 549–583.

27

Anti-Angiogenic Treatment of Endometriosis: Biochemical Aspects

R.N. Taylor[a], M.D. Mueller[b]

[a]Center for Reproductive Sciences, Department of Obstetrics, Gynecology and Reproductive Sciences, University of California, San Francisco, CA, USA; [b]Endometriosis Center, Department of Obstetrics and Gynaecology, Inselspital, University of Bern, Bern, Switzerland

Endometriosis is a common gynecological disorder defined by the presence, growth and progression of endometrial tissue outside the uterine cavity. Endometriosis can be asymptomatic, but it typically is associated with a multitude of clinical symptoms, the most common being pelvic pain and impaired fertility. Studies from our laboratories and several others focus on the neovascularization of endometriosis implants as a critical facet of its pathogenesis. Using the analogy of tumor metastasis [1] we postulated that the angiogenic potential of exfoliated endometrium may predict the likelihood that lesions become established on the peritoneal surface [2]. The growth of new capillary branches from preexisting vessels is complex and requires proteolytic degradation of extracellular matrix, proliferation and migration of endothelial cells, and ultimately the formation of patent capillary tubules supplying the angiogenic stimulus [3]. Recent findings in normal endometrium confirm that vessel growth is maximal during the mid-secretory phase and occurs predominantly via the elongation of capillaries in the subepithelial plexus [4]. While many different growth factors and cytokines have been shown to exert chemotactic and proliferative effects on endothelial cells, vascular endothelial growth factor-A (VEGF) is the most potent of the angiogenic factors. VEGF mRNA was detected in normal and eutopic endometrium of endometriosis subjects by RNase protection analyses, and its expression was found to be highest during the secretory phase in cycling human endometrium [2]. VEGF production also was found to be particularly high in hemorrhagic red implants of endometriosis [5] and in ovarian endometriomas [6]. It was recently reported that the expression of VEGF is higher in cycle phase-matched eutopic endometrium from women with endometriosis than control endometrium [7].

The regulation of bioavailable VEGF is controlled at the transcriptional and post-translational levels. Four or five distinct mRNA species arising via differential splicing of a primary VEGF transcript have been identified and characterized. In the endometrium mRNAs encoding the 165- and 121-amino acid (aa) variants appear to be the predominant isoforms [8, 9], generating glycosylated homodimeric proteins of about 45 and 35 kD, respectively. Longer forms of VEGF (189 and 206 aa) are not actively secreted into the extracellular milieu [10]. Instead, the relatively basic carboxyl termini of these longer isoforms cause them to become reversibly associated with heparan sulfate proteoglycans of the extracellular matrix, where they exert juxtacrine effects.

Transcription of the VEGF gene in primary human endometrial [2] and adenocarcinoma cells [8] is acutely upregulated by estradiol in vitro as it is in rodent uterus in vivo [11, 12]. We mapped a novel estrogen-responsive element (ERE) in the 5′ promoter region of the human VEGF gene [13]. Mutation of the site indicates that this sequence confers responsiveness of endometrial cell VEGF gene expression to estradiol and tamoxifen [14]. The same genetic element was confirmed as the dominant VEGF gene ERE in breast cancer cells [15]. Three canonical progesterone responsive elements (PREs) were identified in the human VEGF gene promoter and progestins directly induced transcriptional activation of the promoter in endometrial cells. However, none of the individual PREs accords a full progesterone receptor-mediated transcriptional response [16]. These new findings indicate that progestational regulation of the VEGF promoter is complex and cannot be localized to confined PRE sequences. It should be emphasized that other factors known to upregulate VEGF gene expression also are very pertinent to the peritoneal environment in endometriosis, including hypoxia, acidosis, prostaglandins and the inflammatory cytokines interleukin-1β and transforming growth factor-β [17–19]. Experiments are in progress to identify the gene sequences that mediate responses to these modulators.

The biomedical implications of defining specific molecular pathways responsible for endometriosis angiogenesis are that they should predict those emerging anti-angiogenic agents that are most likely to be safe and clinically effective for adjunctive therapy in this condition. This is a developing field of study and to date, reports of anti-angiogenic treatment in models of endometriosis have not been published. Nevertheless, several relevant naturally occurring factors are potentially expressed in the endometrium and endometriosis. Moreover, preliminary reports of preclinical therapeutic trials have been presented in abstract form and will be reviewed briefly below.

Endostatin is a carboxy-terminal fragment of collagen XVIII and is a potent angiogenesis inhibitor [20]. Fragments of 16–21 kD in size are found in human plasma [21]. Comparative analyses of collagen expression in normal and ectopic endometrium have not been performed systematically, however, immunohistochemistry and Western blotting failed to detect differences in several extracellular matrix components including collagen IV, laminin, vitronectin, and fibronectin [22]. Another naturally occurring anti-angiogenic factor is the 16-kD amino-terminal fragment of prolactin. Like endostatin, this peptide is generated by enzymatic cleavage of the intact protein and inhibits microvascular proliferation in vivo and in vitro [23]. While it has not been localized to endometriosis per se, the production of active anti-angiogenic 16-kD prolactin fragments has been demonstrated in endometrial stromal cells [24].

Hastings and colleagues reported the use of the soluble flt-1 receptor (VEGF receptor 1) proteins to inhibit estrogen-induced vascularization of endometrium in the mouse uterus [25]. The same group has used this approach to inhibit angiogenesis in nude mice bearing human endometrial transplants. Using the chicken chorioallantoic membrane assay of endometriosis, Nap et al. [26] showed that endostatin, TNP-470 and anti-VEGF antibodies all significantly decreased vascular density within the endometrial transplants.

As new cancer therapy trials move forward into phases II and III, it is likely that agents that inhibit VEGF action, namely naturally occurring VEGF antagonists [27], humanized anti-VEGF antibodies [28], and engineered receptors that trap free VEGF ligands [29] may become available as adjuncts to endometriosis therapy. However, future therapeutic strategies designed to target angiogenic stimuli may not be specific to endometriosis. We must not ignore the likely physiological actions that these same anti-angiogenics will have on eutopic endometrial function, as this also is critically dependent on neovascularization [2]. Indeed, in the short term, anti-angiogenic drugs may be more useful for the treatment of endometriosis-associated pain than infertility.

Acknowledgment

Supported by the NIH/NICHD, through R01-HD33238 and U54-HD37321 as a part of the Special Cooperative Centers Program in Reproduction Research and by a grant from the Swiss National Science Foundation.

References

1 Folkman J: The role of angiogenesis in tumor growth. Semin Cancer Biol 1992;3:65–71.
2 Shifren J, Tseng J, Zaloudek C, et al: Ovarian steroid regulation of vascular endothelial growth factor in the human endometrium: Implications for angiogenesis during the menstrual cycle and in the pathogenesis of endometriosis. J Clin Endocrinol Metab 1996;81:3112–3118.
3 Hanahan D: Signaling vascular morphogenesis and maintenance. Science 1997;277:48–50.
4 Gambino LS, Wreford NG, Bertram JF, et al: Angiogenesis occurs by vessel elongation in proliferative phase human endometrium. Hum Reprod 2002;17:1199–1206.
5 Donnez J, Smoes P, Gillerot S, et al: Vascular endothelial growth factor (VEGF) in endometriosis. Hum Reprod 1998;13:1686–1690.
6 Fasciani A, D'Ambrogio G, Bocci G, et al: High concentrations of the vascular endothelial growth factor and interleukin-8 in ovarian endometriomata. Mol Hum Reprod 2000;6:50–54.
7 Tan XJ, Lang JH, Liu DY, et al: Expression of vascular endothelial growth factor and thrombospondin-1 mRNA in patients with endometriosis. Fertil Steril 2002;78:148–153.

8 Charnock-Jones D, Sharkey A, Rajput-Williams J, et al: Identification and localization of alternately spliced mRNAs for vascular endothelial growth factor in human uterus and estrogen regulation in endometrial carcinoma cell lines. Biol Reprod 1993;48:1120–1128.
9 Huang JC, Liu DY, Dawood MY: The expression of vascular endothelial growth factor isoforms in cultured human endometrial stromal cells and its regulation by 17beta-oestradiol. Mol Hum Reprod 1998;4:603–607.
10 Ferrara N, Houck K, Jakeman L, et al: The vascular endothelial growth factor family of polypeptides. J Cell Biochem 1991;47:211–218.
11 Cullinan-Bove K, Koos RD: Vascular endothelial growth factor/vascular permeability factor expression in the rat uterus: rapid stimulation by estrogen correlates with estrogen-induced increases in uterine capillary permeability and growth. Endocrinology 1993;133:829–837.
12 Shweiki D, Itin A, Neufeld G, et al: Patterns of expression of vascular endothelial growth factor (VEGF) and VEGF receptors in mice suggest a role in hormonally regulated angiogenesis. J Clin Invest 1993;91:2235–2243.
13 Mueller M, Vigne J-L, Minchenko A, et al: Regulation of vascular endothelial growth factor (VEGF) gene transcription by estrogen receptors alpha and beta. Proc Natl Acad Sci USA 2000;97:10972–10977.
14 Mueller MD, Pritts EA, Zaloudek CJ, et al: Regulation of vascular endothelial growth factor (VEGF) by tamoxifen in vitro and in vivo. Gynecol Obstet Invest 2003, in press.
15 Buteau-Lozano H, Ancelin M, Lardeux B, et al: Transcriptional regulation of vascular endothelial growth factor by estradiol and tamoxifen in breast cancer cells: a complex interplay between estrogen receptors alpha and beta. Cancer Res 2002;62:4977–4984.
16 Mueller M, Vigne J-L, Pritts E, et al: Progestins activate vascular endothelial growth factor gene transcription in endometrial adenocarcinoma cells. Fertil Steril 2003;79:386–392.
17 Brogi E, Wu T, Namiki A: Indirect angiogenic cytokines upregulate VEGF and bFGF gene expression in vascular smooth muscle cells, whereas hypoxia upregulates VEGF expression only. Circulation 1994;90:649–652.
18 Ben-Av P, Crofford L, Wilder R: Induction of vascular endothelial growth factor expression in synovial fibroblasts by prostaglandin E and interleukin-1:a potential mechanism for inflammatory angiogenesis. FEBS Letters 1995;372:83–87.
19 Lebovic D, Bentzien F, Chao V, et al: Induction of an angiogenic phenotype in endometriotic stromal cell cultures by interleukin-1beta. Mol Hum Reprod 2000;6:269–275.
20 O'Reilly MS, Boehm T, Shing Y, et al: Endostatin: an endogenous inhibitor of angiogenesis and tumor growth. Cell 1997;88:277–285.
21 John H, Preissner KT, Forssmann WG, et al: Novel glycosylated forms of human plasma endostatin and circulating endostatin-related fragments of collagen XV. Biochemistry 1999;38:10217–10324.
22 Harrington DJ, Lessey BA, Rai V, et al: Tenascin is differentially expressed in endometrium and endometriosis. J Pathol 1999;187:242–248.
23 D'Angelo G, Martini JF, Iiri T, et al: 16K human prolactin inhibits vascular endothelial growth factor-induced activation of Ras in capillary endothelial cells. Mol Endocrinol 1999;13:692–704.
24 Bentzien F, Vigne J-L, Weiner RI, et al: Potential role of cathepsin D in the formation of antiangiogenic fragments of prolactin in the human endometrium. 79th Annual Endocrine Society Meeting Program 1997, pp 1–33.
25 Hastings JM, License D, Comerbach M, et al: The effect of administration of a VEGF antagonist, sflt-1, to oestradiol-treated ovariectomised mice. J Soc Gynecol Invest 2002;9(suppl):608.
26 Nap AW, Groothuis PG, Dunselman GAJ, et al: Anti-angiogenic compounds inhibit angiogenesis and endometriosis-like lesion formation in the chicken chorioallantoic membrane. J Soc Gynecol Invest 2003;10(suppl): 281.
27 Clark DE, Smith SK, He Y, et al: A vascular endothelial growth factor antagonist is produced by the human placenta and released into the maternal circulation. Biol Reprod 1998;59:1540–1548.
28 Ferrara N: Role of vascular endothelial growth factor in physiologic and pathologic angiogenesis: therapeutic implications. Semin Oncol 2002;29: 10–14.
29 Holash J, Davis S, Papadopoulos N, et al: VEGF-Trap: a VEGF blocker with potent antitumor effects. Proc Natl Acad Sci USA 2002;99:11393–11398.

Infertility Issues in the Developing World

28
Infertility: Addressing Challenges in Its Management in Africa

J.K.G. Mati
Institute for Reproductive Health Training and Research, Nairobi, Kenya

Infertility, defined as failure to conceive after a year of unrestricted and unprotected intercourse, is a frequent complaint in gynaecological service in most African countries. The cause of infertility can be found in both male and female partners, both of whom must be investigated in all cases seeking help. Whereas the impact of infertility is felt mainly at the couple and individual levels, in Africa it is also a matter of major concern to the extended family. It is an important problem from the health, social/cultural and economic considerations, one associated with considerable social and mental anguish, not forgetting the heavy financial investment needed in its investigation and treatment.

Magnitude of the Problem

It is estimated that 50–80 million people world-wide are inflicted with infertility, but the magnitude of the problem is greater in the Africa region where it affects 20–30% of the population. Substantial geographic variations are noted, creating the so-called African 'belt of low fertility' that stretches across the continent in a west to east direction, through Togo, Cameroon, Central Africa Republic, Gabon, Demographic Republic of the Congo, to pockets of infertility in parts of Uganda, Kenya, Tanzania and Zambia [1, 2].

The determinants of infertility in the Africa region are numerous, but by far the most important are sexually transmitted infections (STIs) – particularly gonorrhoea and chlamydia, and postpartum/post-abortal infections [3]. As a result, in the Africa region, secondary infertility is encountered more frequently than primary infertility. The prevalence of primary infertility is estimated at less than 5%, while secondary infertility ranges from 10 to 40%.

Causes of Infertility

Ductal Obstruction. The major cause of infertility is ductal obstruction resulting from infections with STIs (particularly chlamydia and gonorrhoea) in both male and female partners, and postpartum/post-abortal infections in the female partner. In the large WHO study on investigation of the infertile couple [4], tubal factor was present in 85% of couples in African centres compared with 36% in centres located in developed countries. Inadequate access to healthcare, often as a result of poor policies and poverty, is the main factor behind the high prevalence of these infections in most African countries. In contrast, ovulation disturbance was detected in 17% of couples, and this often co-existed with tubal disease.

Endometriosis and Infertility. Endometriosis has featured least as a cause of female infertility in sub-Saharan Africa [5]. In a series of 422 laparoscopies in Younde (Cameroon) endometriosis was detected in only 1.6% of the cases [6]. The reasons for the low incidence of endometriosis in African women are yet to be explored. Could it be that the very high frequency of STI-induced PID masks endometriosis? Or is endometriosis being missed because too few laparoscopies are performed? Is it being missed because of lack of training in its identification and because too few biopsies are taken? These are some of the questions that remain to be answered, before we finally accept that the pathology is rare among our populations.

HIV/AIDS and Infertility. The high prevalence of HIV infection in countries with high infertility rates calls for examination of the relationship between the two, if any. HIV infection may decrease fertility through factors related to reduced coital frequency due to ill-health, erectile dysfunction and increased prevalence of STIs among HIV-infected individuals. In males, HIV infection is associated with hypergonadotrophic hypogonadism, with testicular atrophy, spermatogenic arrest and Sertoli-cell-only syndrome [7, 8]. The changes observed in semen parameters increase in severity with progression of the disease. In females, HIV infection exacerbates the severity of pelvic inflammatory disease (PID) thereby causing greater fallopian tube damage.

Tropical Conditions. It is probable that a number of tropical conditions contribute to the low fertility in Africa [9, 10]. Filariasis has been associated with abnormal semen parameters, while tuberculosis and schistosomiasis have been shown to cause tubal obstruction.

Prevention of Infertility

The pattern of infertility in Africa dictates that the leading interventions for reduction of its incidence should be: prevention, early diagnosis and adequate treatment of STIs; improved obstetric care to prevent puerperal sepsis, and prevention and better treatment of unsafe abortion. The WHO/AFRO strategy (1988–2007) for the prevention of infertility has set targets to be achieved in this regard, including the reduction of the prevalence of curable STIs to less than 15% in all countries, and managing effectively at least 80% of cases that seek treatment for curable STIs. In addition the strategy seeks to reduce STIs among adolescents through improved knowledge and health-seeking behaviour.

Management of Infertility

Management of Infertility should best be undertaken at specialised centres, in order to avoid repetition of procedures. Further, in view of the preponderance of ductal obstruction as the cause of infertility in the African setting, it is obvious that this must be excluded early in the investigation of the infertile couple. Thus, the critical steps in the management of infertility may be summarised as follows: (1) demonstration of tubal patency and treatment (as appropriate) of tubal occlusion; (2) demonstration of male fertility and treatment as appropriate; (3) a search for evidence of ovulation and treatment of ovulation failure; (4) immunological investigation involving both partners.

Several constraints can be identified that interfere with effective management of infertility in Africa, some of these are lack of human and material resources, late presentation of cases, the gross pathology commonly encountered during investigation, illiteracy and poverty. The results of surgical treatment of tubal disease are generally poor mainly because many of the cases present late with a lot of adhesions, falling in Classes 2 and 3 of tubal disease [11, 12, 5]. In fact only 12% of the cases investigated in Nairobi were considered suitable for reconstructive surgery [13].

One of the ironies of nature is that assisted reproductive technology (ART), originally developed to overcome infertility due to tubal factor, today remains largely inaccessible in the continent that may rightly be described as the 'mother of all tubal infertility'! Whereas prevention is always better than cure, society cannot forget the plight of millions already affected by the problem, and certainly the preventive measures will never alleviate the suffering of the affected. One can read a parallel to the recent 'change of heart' with regard to increasing access to anti-retroviral drugs (ARV) in poor nations.

Other than in South Africa, ART facilities are limited in the sub-Saharan Africa, and it is only in a few countries (Cameroon, Ghana, Nigeria, Senegal, Togo and Zimbabwe) that serious attempts have been made in this direction, almost all in the private sector (Giwa-Osagie, personal communication, 2003). There are two main reasons why this procedure has not been instituted in public teaching hospitals. The first is the high investment cost needed to set up the facilities, but perhaps the main obstacle has been attitudes of governments and doctors, who feel there are more 'important' health problems to spend meagre resources on, other than childlessness.

Summarised below are areas that need to be considered in strategic planning for increased access to ART in the Africa region:

Advocacy: There is need for advocacy aimed to make governments to recognise importance of infertility. This will be more effective if it is backed by some data on the economic impacts of infertility, in terms of its magnitude, the social and economic impacts to the couple, its potential effects on productivity, and the costs involved in its treatment abroad including drain in foreign exchange. Governments need to be urged to increase resource allocation for the prevention and treatment of infertility.

Training: there is need to improve the quality of training on the management of infertility. This is particularly important in centres with ongoing residency training. This may also involve attraction of external (local or international) resources for the creation of regional centres of excellence, which can provide training for groups of countries, as well as undertaking research.

Technical exchange: the quality of training can be enriched through technical exchange and assistance, both in the North-South as well as in the South-South directions.

Research needs to be intensified towards simplifying ART technology in order to make it more affordable and robust, to be applied in areas of limited resources.

The role of AIH and AID should not be ignored in preference for the more 'flushy' technological approaches, and neither must adoption be forgotten. Adoption can provide a couple with the much sought-after child; it opens an entirely new world and opportunities for the adopted child.

References

1 Belsey M: The epidemiology of infertility: A review with particular reference to sub-Saharan Africa. Bull WHO 1976;54:319–341.
2 Larsen U: Primary and secondary infertility in sub-Saharan Africa. Int J Epidemiol 2000;29:285–291.
3 World Health Organisation: Infections, pregnancies and infertility: Perspectives on prevention. Fertil Steril 1987;47:964–968.
4 Cates W, Farley TMM, Rowe PJ: Worldwide patterns of infertility: Is Africa different? Lancet 1985;14:596–598.
5 Nasah BT, Otubu JAM: Infertility; in Contemporary Issues in Maternal Health Care in Africa. Harwood Academic Publishers, 1994, pp 337–370.
6 Doh AS, Formulu JN, Leke RJ, Nasah BT: A five year retrospective review of laparoscopy in Yaounde. Ann Universit Sci Santé 1986;30:166–171.
7 Busingye BR, Sekadde Kigondu C, Wango EO, Mmiro FA: Reproductive hormones and testicular histology in males with human immunodeficiency virus infection. 1st Pan African Conference on Biochemistry and Molecular Biology 1996, p 279.

8 Dobs AS, Dempsey MA, Ladenson PW, Polk BF: Endocrine disorders in men infected with Human Immunodeficiency Virus. Am J Med 1988;84: 611–616.
9 Nasah BT and Cox JN: Vascular lesions associated with infertility in Cameroon: Possible relation to parasitic disease. Virchow Archiv A, Pathol Anat Histopathol 1979;377:225–236.
10 Rogo KO, Sekadde Kigondu CB, Muitta MN, Njoroge JK, Mati JKG: The effect of tropical conditions on male fertility indices. J Obstet Gynaecol Eastern Centr Africa 1985;4:45–52.
11 Rock JA, Katayama KP, Martin EJ, Woodruff JD, Jones HW Jr: Factors influencing the siccess of salpingostomy techniques for distal fimbrial obstruction. Obstet Gynecol 1978;52:591–596.
12 Capsi E, Halperin Y, Bukovsky I: The importance of periadnexal adhesions in tubal reconstructive surgery for infertility. Fertil Steril 1979;31:296–300.
13 Mati JKG: The pattern of infertility in Kenya. Proceedings of the Third European Congress on Sterility, Athens, Greece, 1972, pp 274–278.

29

The Need for Infertility Services in the Developing World: WHO Point of View

O.F. Giwa-Osagie

Department of Obstetric Gynecology, College of Medicine, University of Lagos, Lagos, Nigeria

The definition of 'Health' by the WHO, by implication, and its definition of 'Reproductive Health' include management of infertility. The uptake of infertility management services in developed countries and the prevalence of causes of infertility in developing countries [1–3] highlight the need for infertility services in developing countries. Infertility accounts for 40–60% of all gynaecological consultations in developing countries. Infertility is a cause of marital discord, great unhappiness and even suicide in developing countries. Developing countries have diminishing resources while developed countries are increasingly exhibiting donor fatigue. What resources are available from donors is being competed for by major health issues such as HIV/AIDS, malaria, immunisation, nutrition and gender issues [4]. It is therefore important that the international community and the individual national communities must be aware of the need to provide resources for infertility prevention and management. Prevention of STIs, unwanted pregnancies, unsafe abortion and deliveries in unhygienic environment and without trained supervision will drastically reduce the prevalence of infertility and the distribution of causes of infertility. In addition parts of the prevention of infertility crosscut with other health priorities such as prevention of STIs, HIV/AIDS, adolescent reproductive health, contraception, safe motherhood and making pregnancy safer (MPS). These crosscutting issues should be used to access funding for infertility management.

Rational, cost-effective investigation of infertility has to be the bedrock for the management of infertility in the developing world. The adoption of an algorithm for infertility management with regional and country adaptations will streamline infertility management at the primary, secondary and tertiary levels of healthcare. The more detailed and expensive investigations and treatments will be available at selected referral centres. Algorithms for management of infertility are available in various WHO Regions and need to be used more widely. Enzyme-linked immunoassays for reproductive hormones have virtually replaced radio-immunoassays and these methods are cheaper and more appropriate for developing countries. In spite of the high cost of starting and offering services in advanced infertility management such as assisted conception these methods are now available in several countries in the developing world [5, 6]. The WHO should facilitate prevention of infertility strategies, encourage and enhance the development of appropriate cost-effective investigations and management of infertility, facilitate capacity building in applicable technology and provide for enabling funding for collaboration between centres in developing world and between them and appropriate units or centres in developed countries.

References

1 Giwa-Osagie OF, Ogunyemi O, Emuveyan EE, Akinla O: Aetiologic classification and socio-cultural characteristics of infertility in 250 couples. Int J Fertil 1984;29:104–108.
2 Cates W, Farley TMM, Rowe PJ: Worldwide patterns of infertility: Is Africa different? Lancet 1985;i:596–598.
3 Yeboah ED, et al: Aetiological factors of male infertility in Africa. Int J Fertil 1987;37:300–307.
4 Buve A, Bishikwabo-Nsarhaza K, Mutangadura G: The spread and effect of HIV-1 infection in sub-Saharan Africa. Lancet 2002;359:2011–2017.
5 Ogedengbe OK, Giwa-Sagie OF: Implications of pattern of tubal disease for microsurgery and in vitro fertilisation in Lagos. J Na Med Ass 1987;75: 510–513.
6 WHO: International Conference on Assisted Reproductive Technology, Geneva, September 2001.

30

Perception of Infertility in Two Communities in Kenya

C. Sekadde-Kigondu[a], V.N. Kimani[b], L.W. Kirumbi[c], J.K. Ruminjo[b], J. Olenja[d]

Departments of [a] Clinical Chemistry, [b] Community Health, [c] Obstetrics and Gynecology, University of Nairobi, and [d] Kenya Medical Research Institute, Nairobi, Kenya

Introduction

Infertility is defined as the failure to conceive after a year of regular coitus without any form of contraception. The couple must be considered as a unit when investing their condition. In Kenya, the major cause of infertility in women is as a consequence of sexually transmitted disease (STI) or unsafe abortion leading to tubal occlusion. This effect of STIs in men cannot be underscored as well. It is well known that STIs are preventable but this knowledge was lacking in the communities.

Objectives were to: (1) assess the communities' knowledge on causes and management of infertility; (2) assess the communities' knowledge on the preventive aspects of infertility; (3) assess their recommendations on the prevention infertility; (4) make recommendations.

Study Design

Descriptive Study

Study Population. Communities in the rural and urban areas (adolescents, married men and women, infertile men and women and providers).

Study Areas. Low income area in Nairobi, the capital city of Kenya and a rural community in Muranga. The populations had contrasting prevalence of STIs, higher in the urban area than the rural area.

Study Methods. Data was collected using qualitative methods (focus group discussion and in-depth interviews) with study guides.

Results

Definition of infertility was according to the cultural backgrounds but differed from the medical definition since many communities would wait for a long time even six years before they sought for medical care. The infertile women were assigned special names, which created a gender bias since the man was ignored in this respect.

The communities had poor knowledge on the causes of infertility and only a few had an insight in the role of the STIs in the causes of infertility. Religion played an important part as one of the major factors leading to infertility. Observing religious ceremonies was deemed to reverse infertility.

Medically acknowledged causes of infertility in both males and females were elusive to the common man and woman including those who have been affected. The causes of infertility were gender specific and centered around the ability of the female to produce 'eggs' or the man to produce the 'seeds' (spermatozoa). The heat was associated with non-viable eggs if the woman had too little 'heat'. Blood incompatibility was also attributed to the causes of infertility in the couple, as was non-performance (impotence) in man due to misuse of drugs. In many cases the woman was blamed for the condition.

Management of infertility was based on removing curses and seeking formal medical care was always delayed and deemed to be expensive for many of the infertile women. In many cases they moved from one health provider to another without positive result. This is understandable since 67% of infertile women presenting at Kenyatta National Hospital in Kenya were diagnosed with a tubal factor. The providers also had poor knowledge due to low interest in its management.

Although the condom is known to prevent STIs there was little effort to use it in the two communities. Preventive measures of infertility were centered on increasing knowledge especially in the youth.

Conclusion

The study revealed that infertility is ignored as a health problem in the two communities. The formal and informal management of infertility was not coordinated and in some cases too expensive to be utilized by the infertile couples. Effort should be made to update the providers' knowledge on the management of infertility and to encourage the patients to seek formal medical care at the earliest possible chance. There is need to educate the youth in the causes of infertility and its preventive aspects. Low-cost assisted reproductive technology should be established so that the infertile couples in Kenya can benefit from it.

Gynecologic and
Obstetric Investigation

Author Index Vol. 57, No. 1, 2004

Numbers refer to abstract numbers

KARGER

Fax + 41 61 306 12 34
E-Mail karger@karger.ch
www.karger.com

Accessible online at:
www.karger.com/goi